Instabilities and
Catastrophes in
Science and Engineering

A thin elastic spherical shell buckling due to a uniform external pressure.
Computer graphics here and elsewhere by Richard J. Thompson

Instabilities and Catastrophes in Science and Engineering

J. M. T. Thompson

Professor of Structural Mechanics
University College London

A Wiley–Interscience Publication

JOHN WILEY & SONS

Chichester · New York · Brisbane · Toronto · Singapore

British Library Cataloguing in Publication Data:
Thompson, J. M. T.
  Instabilities and catastrophes in science
  and engineering.
  1. Structures, Theory of
  1. Title
  624. 1'7   TA645

  ISBN 0 471 09973 2 (cloth)
        0 471 10071 4 (paper)

Photosetting by Thomson Press (India) Ltd., New Delhi and
printed at the Pitman Press Limited, Bath, Avon.

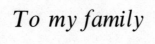

To my family

Contents

Foreword

There is one great complicating feature that introduces major difficulties into mechanics, physics, chemistry, engineering, astronomy, and biology. This complicating feature is that an equilibrium can be stable but may become unstable; while, similarly, a process can take place continuously but may become discontinuous.

In recent years an impressive collaboration between workers in all of the above disciplines and applied mathematicians has brought into being a marvellous body of general theory giving a most valuably informative overall view of these difficulties. Yet that body of theory has frequently been expounded in a style that is much too technically intricate to be widely accessible.

Nobody could have been better qualified than Professor J. M. T. Thompson to prepare the present splendidly clear account of the entire subject. It can be recommended most warmly to every reader interested in important modern developments in the sciences or in engineering.

SIR JAMES LIGHTHILL

Preface

Instability phenomena are of vital concern to all scientists and engineers, and this informal book presents a personal pictorial survey of some modern, interesting, and important examples drawn from the breadth of science and technology. They include the collapse and galloping of slender structures under gravitational and wind loading, the exotic astrophysics of collapsing stars, the unexpected fracture of a crystal lattice, the thermodynamic self-organization of biochemical systems, the population explosions of competing ecological species, the onset of turbulence in a fast-moving fluid, the recently discovered chaotic motions of simple deterministic models, the flutter of high-speed aircraft, the attitude control of spacecraft, and the neural dynamics of the brain.

A unified view of all these diverse instabilities is achieved by drawing on new ideas of bifurcation and catastrophe theory, but the presentation is held, as far as possible, at the level of popular magazines such as *Scientific American* or *New Scientist*, and is aimed at a similar wide general audience. One very important feature is the extensive list of up-to-date research references, numbering 337, which can lead an interested reader quickly into the specialist technical literature of any topic.

To understand the subject matter in any depth, *some* mathematics is essential, and the reader is introduced gently and systematically throughout the book to many of the mathematical preliminaries, always staying however well within the normal background of most engineers and scientists. First-year university students in any branch of science or technology should follow the text with ease, an elementary knowledge of algebra, the calculus, and simple differential equations being all that is assumed.

The mathematical theory of bifurcations and instabilities, with its historical origins deep in mechanics and astrophysics, has indeed undergone spectacular developments in recent years, especially at the hands of the topologists. Two particular outcomes have been the deep *catastrophe theory* classifications of Thom and Zeeman, grounded in the topological concept of *structural stability*, and the current excitements about the seemingly chaotic motions of *strange attractors*.

The bulk of this modern theory, written often within the notations and concepts of abstract topology, is not readily accessible to many interested research workers versed in conventional applied mathematics, and it is hoped that this small book will serve to introduce these workers to some of the general ideas involved, while pointing the way through the references to higher things.

The book delineates both the *static* instabilities that have been the main ingredient of catastrophe theory and also the *dynamic* instabilities that arise, for example, in the wind-induced flutter of aircraft and suspension bridges, where the response is not governed by an energy potential. Home-made games and experiments are described to encourage the reader to try things out for himself, and the book contains about 150 detailed diagrams and photographs.

Much of the underlying theory is to be published shortly by John Wiley in a companion book entitled *Static and Dynamic Instability Phenomena*, by J. M. T. Thompson and G. W. Hunt, and I am indeed deeply indebted to my colleague Giles Hunt for his continuing collaboration on many topics.

Chapter 1, a general introduction to this volume, is based on my Inaugural Lecture 'Instabilities and Catastrophes in the Physical Sciences' delivered at University College London on the third of May, 1979. It is reproduced by courtesy of University College and the *Journal of Engineering Sciences* of the University of Riyadh where it was reprinted as 'Static and Dynamic Instabilities in the Physical Sciences'.[274]

Section 10.2 was drafted by Professor Tom Kane of Stanford University based on a recent article of his with D. A. Levinson,[327] and I am most grateful to these two researchers for this contribution.

J. M. T. THOMPSON

CHAPTER 1

Introduction

We give in this introductory chapter a brief survey of some of the stability concepts and phenomena that we shall be meeting in greater detail in the body of the book.

1.1 Historical survey from Newton to Andronov

Sir Isaac Newton of course laid the foundations of mechanics[1] in his *Principia* of 1686 and it is interesting to find in his work detailed experimental studies of the motions of a simple pendulum in both air and water. The damped oscillations of such a pendulum are an archetypal example of an asymptotically stable system, and his results remind us that air is by no means a linear viscous damper.

Some fifty years later in 1744 the famous Swiss mathematician Leonard Euler used his newly invented calculus of variations to determine the equilibrium configurations of a compressed elastic column.[2] This first study of a *bifurcation* problem is commemorated in engineering mechanics in the terminology of an 'Euler strut'.

Lagrange, a young acquaintance of Euler, developed the *analytical* energy approach to mechanics which allows important generalizations that cannot easily be made in the Newtonian vector approach.[3] This led naturally to his fundamental energy theorem that a minimum of the total potential energy is sufficient for stability.

Analytical mechanics was further developed by Hamilton, who emphasized the vector field of the phase trajectories by the reduction to a set of first-order differential equations.

The founding father of bifurcation theory as we know it today was the outstanding French mathematician Henri Poincaré. He shared with Leonard Euler a prodigious memory and the ability to perform vast tracts of analysis entirely in his head amidst the noisiest distractions. Within his total output of about 500 papers and over 30 books covering the whole of the mathematics of his day, Poincaré sketched a general bifurcation theory[4-6] and created the global qualitative dynamics in which much of stability theory must now be viewed.

Brief biographies of Euler, Lagrange, and Poincaré are given in an Appendix to this book.

Liapunov gave mathematical precision to the basic definitions of stability, and introduced the generalized energy functions that bear his name, in his definitive memoir[7] of 1892.

1

Following in the path of Poincaré, Andronov and Pontryagin[8] introduced in 1937 the important topological concept of *structural stability* that underlies the current classifications of Thom, Zeeman, Smale, and Arnold.[9-15] This blossoming of Poincaré's qualitative dynamics now represents a major topological contribution to the foundations of mechanics in general and of stability in particular, as can be judged by the second edition of the recent monograph of Abraham and Marsden.[16]

This historical sketch of stability theory grounded in classical mechanics is summarized in Figure 1.

The exponential growth of science and applied mechanics has led to the fragmentation and diversification of the original classical development, but we will here sketch briefly the specialized field of *elastic stability*. This is concerned with the response of elastic solids and structures to some form of mechanical loading, and has important technical applications in assessing the buckling strength of engineering structures. A useful distinction can be drawn here between conservative systems on the one hand (including in this category systems with small energy dissipation which are otherwise conservative) and non-

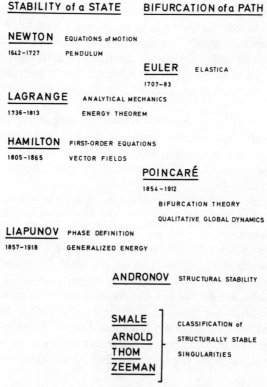

STABILITY of a STATE   BIFURCATION of a PATH

NEWTON   EQUATIONS of MOTION
1642–1727   PENDULUM

EULER   ELASTICA
1707–83

LAGRANGE   ANALYTICAL MECHANICS
1736–1813   ENERGY THEOREM

HAMILTON   FIRST-ORDER EQUATIONS
1805–1865   VECTOR FIELDS

POINCARÉ
1854–1912

BIFURCATION THEORY

QUALITATIVE GLOBAL DYNAMICS

LIAPUNOV   PHASE DEFINITION
1857–1918   GENERALIZED ENERGY

ANDRONOV   STRUCTURAL STABILITY

SMALE
ARNOLD   CLASSIFICATION of
THOM   STRUCTURALLY STABLE
ZEEMAN   SINGULARITIES

Figure 1   Historical development of basic stability theory

conservative systems on the other, these normally having an unlimited energy source.

In the former conservative category we must first mention the classic bifurcation studies of Koiter[17] in his definitive dissertation at Delft in 1945. His study was grounded in a continuum formulation, but an elimination of passive deformations yielded a final algebraic energy function of the buckling mode amplitudes, a manoeuvre that lends validity to our subsequent study of discrete systems. A modern account of the non-linear branching of continuous elastic structures under conservative loading is given by Budiansky,[18] and in the same volume of *Advances in Applied Mechanics* can be found Hutchinson's important extension to the instability of structures loaded into the plastic range.[19]

In the non-conservative field, important linear classifications were made by Ziegler[20] in 1956 and 1968, and the extensive excellent work of Herrmann,[21] Leipholz,[22-25] and their colleagues should also be mentioned. These non-conservative elastic studies are predominantly confined to the *linear* small-deflection range.

Work in the Stability Research Group at University College London has been concerned primarily with the *non-linear* branching of *discrete* or discretized conservative systems, and some significant publications are listed.[26-35] This group was initiated in the nineteen-sixties by Sir Henry Chilver, now Vice Chancellor of the Cranfield Institute of Technology, and its work has been closely linked with the finite-element analysis of engineering structures. Much of its research is presented in the 1973 monograph of Thompson and Hunt,[36] and more recent work embracing catastrophe theory and non-conservative problems is to be presented in a current monograph by the same authors.[37] This latter monograph can be seen as a companion volume to this present book, laying the analytical foundations of our following discussions.

A significant linkage with the earlier chart of Figure 1 is provided by our recent interaction with the catastrophe theory of René Thom and Christopher Zeeman.[9-11] The deep and stimulating parallels that have emerged between the engineering and topological approaches have been described in a number of papers[34,38] and seem to represent a useful step towards a unified static bifurcation theory.[39]

1.2 Instabilities of a linear oscillator

Leaving this brief historical survey, we now outline some of the basic ideas of stability theory before moving on to outline the more recent ideas underlying this book.

The distinctive *static* and *dynamic* instabilities are nicely illustrated with reference to the damped linear oscillator of Figure 2. Here a mass m is restrained by an elastic spring of stiffness s and a dash-pot damper which provides a viscous force opposing the velocity. Plotting the displacement x against the time t we have the familiar damped oscillatory motion typical of a pendulum vibrating in

4

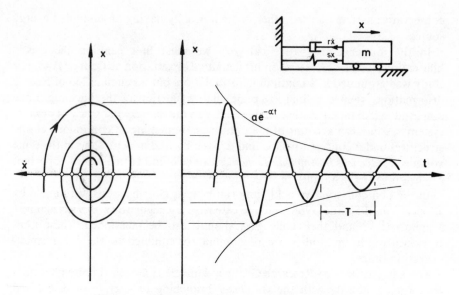

Figure 2   The response of a damped linear oscillator

air. In the *phase space* of x against $\dot{x} = dx/dt$ we have a stable focus replacing the familiar circles or ellipses of an undamped system. A three-dimensional graph of this *asymptotically stable* behaviour is shown in Figure 3, in the space of $x, \dot{x}$, and t.

The equation of motion of this oscillator is shown in Figure 4 and dividing by the mass puts this in convenient standard form. Looking for an exponential solution leads to the boxed *characteristic* equation which is here simply a quadratic in λ.

The form of solution depends on the roots of this characteristic equation, which in turn depend on the sign of the discriminant D. If D is positive, we have two real roots (solid circles) and the assumed exponential solution, while if D is negative we have two complex conjugate roots (open circles) giving us solutions of the form $e^{Rt} \sin It$. The damped oscillator therefore becomes unstable if any root acquires a *positive real part*.

The response of the oscillator is summarized in Figure 5. If the stiffness is large and the damping small the roots are complex as indicated, and we have the stable focus of our earlier discussion. If we decrease the stiffness following the horizontal arrow the roots become real as we cross the parabola of critical damping, $D = 0$, and the phase portrait becomes a stable node.

Now a load on an elastic structure can induce static *buckling* in which the effective stiffness of the system changes from positive to negative. This *static instability*, characterized by the appearance of an adjacent position of equilibrium, is represented by the horizontal arrow. If, on the other hand, a wind blows across a flexible elastic structure it can induce dynamic *galloping* in which the effective damping becomes negative, as indicated by the vertical arrow. In this

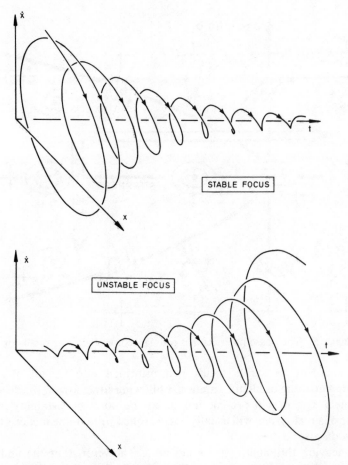

Figure 3 Phase-time trajectories for a stable and an unstable focus

<u>LINEAR OSCILLATOR</u>

$$\underline{m\ddot{x} + r\dot{x} + sx = 0}$$
$$\underline{\ddot{x} + b\dot{x} + cx = 0}$$

$$\underline{x = e^{\lambda t}}$$

$$\boxed{\lambda^2 + b\lambda + c = 0}$$

$$\underline{D = b^2 - 4c}$$

$D > 0 \longrightarrow \lambda = R_1, R_2$
$D < 0 \longrightarrow \lambda = R \pm Ii$

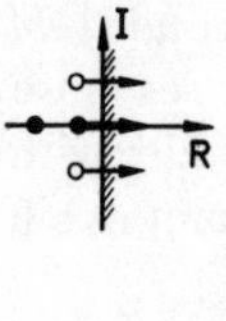

Figure 4 Equation of motion and characteristic roots of a linear oscillator

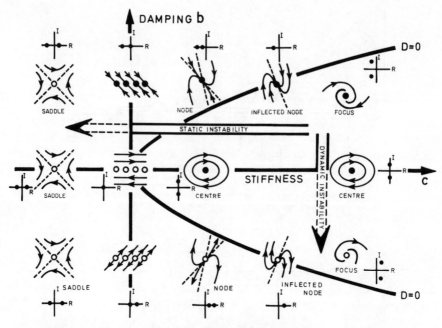

Figure 5 The phase portraits and root structure of a linear oscillator

dynamic instability a stable focus is transformed into an unstable focus representing growing oscillatory motion. Following either arrow, the deflections of our linear system will become infinite at the point of instability, but the behaviour of a real system will usually be controlled by non-linear effects that we shall now discuss.

Before leaving this figure we should notice, however, that the undamped conservative system with a centre of elliptical paths is really a critical marginal case between stable and unstable zones, and the discussion of *elastic stability* without damping gives the pathological routes of Figure 6. This is often forgotten, and the neglection of damping, which is safe for an otherwise conservative system, can lead to paradoxical results for gyroscopic and circulatory systems as demonstrated by Herrmann[40] and Ziegler.[41] We shall examine this in detail later.

1.3 Non-linear static and dynamic bifurcations

Turning to the role of non-linearities, three common static bifurcations are shown in Figure 7, which models a structure as a ball rolling on an energy surface which deforms as a load Λ is applied. In the first *asymmetric* point of bifurcation,[36] familiar in the buckling of engineering frames, a maximum and a minimum coalesce and then re-emerge. In the second *stable-symmetric* point of bifurcation, familiar to structural engineers in the response of an Euler column, an original minimum is transformed into a broad valley with a small central peak. The third picture shows the *unstable-symmetric* point of bifurcation which is the

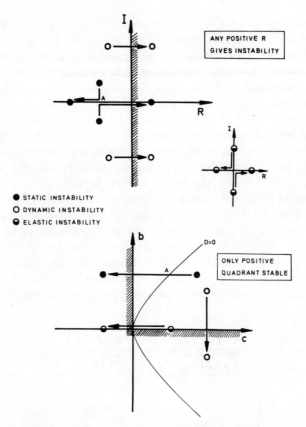

ANY POSITIVE R
GIVES INSTABILITY

● STATIC INSTABILITY
○ DYNAMIC INSTABILITY
◒ ELASTIC INSTABILITY

ONLY POSITIVE
QUADRANT STABLE

Figure 6   Summary of the instability mechanisms

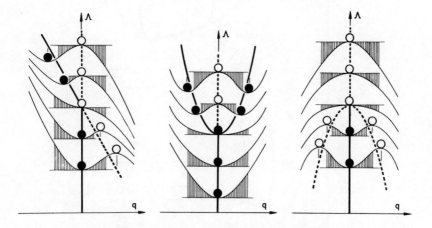

Figure 7   Three familiar static bifurcations

inverse of the previous case, and is common in the buckling of arches and shells.

In each of these three bifurcations we see that the trivial equilibrium state of zero deflection q becomes unstable on intersecting a secondary equilibrium path. This corresponds to a recently established basic theorem[36,42,43] that guarantees that for a conservative system an instability will always be signalled by a non-linear bifurcation, in the absence of a limit point.

Dynamic points of bifurcation are perhaps a little less familiar, and a simple example of a dynamic bifurcation is shown in Figure 8, relating to the displayed non-linear oscillator. Here, in place of a symmetric post-critical *equilibrium path* we have a trace of growing *limit cycles*, which are viewed here in the phase space x against $\dot{x}$ at various fixed values of the loading parameter Λ.

For D positive we have the dynamic analogue of the stable-symmetric point of bifurcation. The trivial state $x = 0$ is stable for Λ less than Λ^C and we have an attracting focus. For Λ greater than Λ^C this state corresponds to an unstable repelling focus and all local motions of the system tend to the stable attracting limit cycle representing a stable finite maintained oscillation. The size of this limit cycle increases with Λ from zero at the critical equilibrium state.

When the non-linear coefficient D is negative, we have the dynamic analogue of

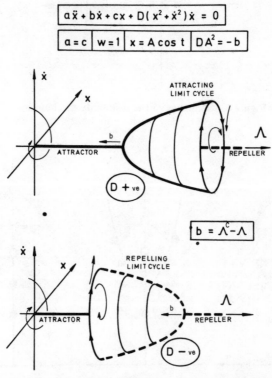

$$a\ddot{x} + b\dot{x} + cx + D(x^2 + \dot{x}^2)\dot{x} = 0$$

| $a = c$ | $w = 1$ | $x = A\cos t$ | $DA^2 = -b$ |

$$b = \Lambda^c - \Lambda$$

Figure 8 Dynamic bifurcation of a non-linear oscillator

the unstable-symmetric point of bifurcation. The trivial equilibrium solution again loses its stability at $\Lambda = \Lambda^C$ at which an attracting focus becomes a repelling focus. The stable sub-critical equilibrium state has, however, only a finite domain of attraction bounded by a repelling unstable limit cycle. A finite disturbance that carries the system beyond this cycle would give rise to growing oscillations even though the control parameter Λ is less than its critical value. The stable trivial solution is here described as *meta-stable*, as in the corresponding unstable-symmetric static bifurcation. The most typical dynamic bifurcations of this type are called Hopf bifurcations, after a celebrated theorem.[44,45]

It is perhaps instructive to summarize the phase transitions of the four symmetric bifurcations on one diagram, and this is done in Figure 9. Once again phase portraits are sketched at sub- and super-critical loading, and the analogies between the two static and the two dynamic bifurcations are nicely displayed.

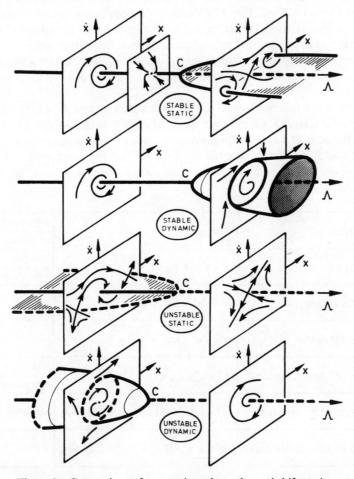

Figure 9   Comparison of two static and two dynamic bifurcations

Notice, however, that for the two static bifurcations, a focus is always replaced by a node close to the bifurcation: this is not always specifically drawn.

In hydrodynamics, stable/unstable-symmetric points of bifurcation are called super/sub-critical bifurcations, and the topologists often call a symmetric bifurcation a pitchfork, for obvious reasons. The asymmetric point of bifurcation is called a trans-critical bifurcation.

We shall return to the dynamic bifurcations later when we discuss the fluid loading of structures, but we turn now to a closer examination of the static bifurcations that are typically observed in the response of conservative systems.

1.4 Static bifurcations and catastrophe theory

An essential aid to the classification of static instabilities is provided by catastrophe theory, a basic table of which is shown as Figure 10.

To establish this table for discrete conservative systems governed by a potential function, René Thom invoked the topological concept of *structural stability* to argue that the experimentally *observable* forms of instability depend on the number of operational control parameters. Thus in our table of the seven elementary catastrophes, if we have only a single control parameter λ we can only *typically* observe the fold catastrophe which has a local potential function of the form shown.

If we have *independent control* of two parameters λ^1 and λ^2, which might for example be a lateral load and an axial load on a column, we can additionally observe the cusp. If we have independent control of three parameters λ^1, λ^2, and λ^3 we can additionally observe the swallow-tail and the hyperbolic and elliptic umbilic catastrophes, while with four controls we can observe any catastrophe on our list.

FOLD	$\underline{q^3} + \lambda q$	LIMIT POINT
		ASYMMETRIC
CUSP	$\underline{q^4} + \lambda^2 q^2 + \lambda^1 q$	STABLE-SYM
		UNSTABLE-SYM
SWALLOW-TAIL	$\underline{q^5} + \lambda^3 q^3 + \lambda^2 q^2 + \lambda^1 q$	
BUTTERFLY	$\underline{q^6} + \lambda^4 q^4 + \lambda^3 q^3 + \lambda^2 q^2 + \lambda^1 q$	
HYPERBOLIC UMBILIC	$\underline{q_2^3 + q_1^3} + \lambda^1 q_2 q_1 - \lambda^2 q_2 - \lambda^3 q_1$	MONOCLINAL
		HOMEOCLINAL
ELLIPTIC UMBILIC	$\underline{q_2^3 - 3 q_2 q_1^2} + \lambda^1 (q_2^2 + q_1^2) - \lambda^2 q_2 - \lambda^3 q_1$	ANTICLINAL
PARABOLIC UMBILIC	$\underline{q_2^2 q_1 + q_1^4} + \lambda^1 q_2^2 + \lambda^2 q_1^2 - \lambda^3 q_2 - \lambda^4 q_1$	

Figure 10   List of the seven elementary catastrophes

This list, then, includes all the structurally stable singularities that can be observed in the real world with the operation of up to four independent control parameters. It has immediate relevance in developmental biology, where space and time are often the ultimate control parameters of differentiating cells, as we shall see later in the book. We should note here that the first four catastrophes in our list have only a single *active* generalized coordinate, as in simple buckling, q, while the last three have two active coordinates, as in simultaneous buckling, q_1 and q_2.

On the right-hand side we have given our engineering terminology for some of these singularities, and we shall see later that the *finer* sub-classification, which is not meant to be complete, arises from our concentration on a single distinctive control parameter.

Essentially Thom is saying that we can only observe a very pathological situation if we have a sufficiently high level of control, and we demonstrate this now for the fold catastrophe following an earlier exposition.[46]

1.5 The fold or limit point

As we have seen, the fold has only a single active coordinate and we can illustrate Thom's argument by reference to Figure 11. Here the total variables, capital Q and Λ, replace the incremental variables that we have been writing as lower case q and λ, and V is the total potential energy.

Now if a child scribbles on the blackboard a form for V as a function of Q, his curve will typically exhibit maxima and minima at which $\partial V/\partial Q$ is zero, but we cannot expect a horizontal point of inflection at which the first and second derivatives vanish simultaneously. The occurrence of such a *critical* point would be considered pathological, and we could assign to it a probability of zero. In complete analogy with the scribble on the blackboard, if we build a model structure in a laboratory and apply a fixed load, we would not expect to find the structure exactly at a critical equilibrium state.

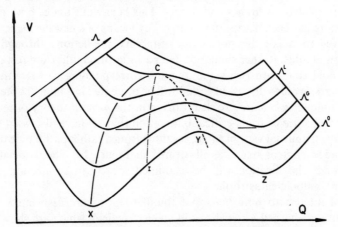

Figure 11 The energy transformation in the fold catastrophe

12

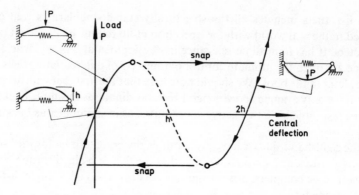

Figure 12   The load-deflection behaviour of a shallow tied arch

The only way to observe a horizontal point of inflexion on the blackboard is to draw a *family* of curves parametrized by Λ as we have done in the figure. Thus to observe a critical equilibrium state at which the first two derivatives vanish simultaneously we must make a parameter scan through one Λ, and this is exactly what we do when we load a structure until it snap-buckles at a *fold* or *limit point*.

This most simple energy transformation, involving the coalescence and extinction of a minimum and a maximum under the operation of a single control is termed a *fold catastrophe*. It yields an equilibrium path XCY which folds over and changes its stability at the critical point C.

Folds are familiar throughout science and technology, and perhaps the simplest mechanical illustration is provided by the *tied arch* of Figure 12. Here an originally straight springy metal strip of steel is held in the form of a shallow arch by a horizontal elastic spring. On loading with a dead hanging mass at the crown, this arch can jump through into the inverted position as indicated, performing a *hysteresis cycle* as shown under slow cyclic loading.

Figure 13 shows the more complex folding behaviour that can arise in the large deflections of shallow arches and domes. This behaviour under both dead and rigid loading has been discussed recently[47] by means of a new *conjugate theorem* that serves to make the not-obvious stability predictions through such a succession of folds. Rather similar folded paths in the equilibrium response of a massive cold star, with two stable regimes corresponding to *white dwarf* and *neutron* stars, are also discussed,[47] and will be presented in a later chapter. This is a rather unique *mechanics* problem arising in the *general* theory of relativity.

The asymmetric or trans-critical bifurcation that we have discussed earlier is essentially an unusual view of a fold, resulting from a rather pathological route in an expanded control space, as illustrated in Figure 14. The associated finer bifurcational classification will be examined shortly in connection with the hyperbolic umbilic catastrophe.

It is of interest to note here that thunder-storms are associated with the instabilities of charged water drops in an electric field, which are shown by the analyses of Rayleigh and Taylor[48] to be generated by just such a trans-critical

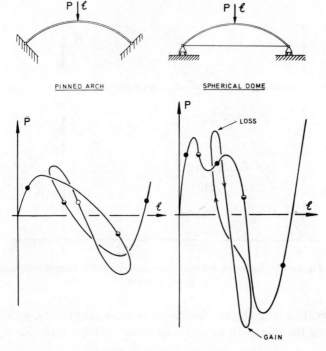

Figure 13   Successive folds in the equilibrium paths of a shallow
arch and a spherical dome

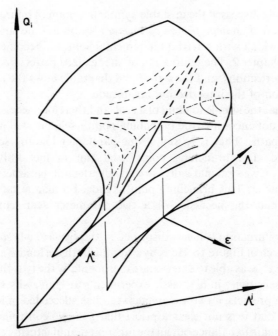

Figure 14   Diagram showing the asymmetric bifurcation as a fold catastrophe

14

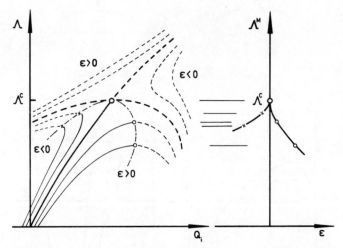

Figure 15 Imperfection sensitivity in an unstable-symmetric point
of bifurcation

bifurcation. This is an elegant illustration of branching theory, with the electric field playing the role of an initial imperfection destroying the basic symmetry.

1.6 The cusp or symmetric bifurcation

We have already discussed the unstable-symmetric point of bifurcation, and with the introduction of an *imperfection parameter* ε we obtain the familiar[36] pictures of Figure 15, which with a trivial fundamental solution becomes equivalent to Figure 49 of chapter 2. The bifurcation of an idealized perfect system is rounded off by the imperfections, and the right-hand diagram shows the failure load as a cusped function of the imperfection magnitude.

It is clear that the simple loading of a *real*, and therefore necessarily imperfect, structure will not encounter the branching point itself, but will just yield one of the rounded paths. That is to say, a one-parameter loading scan will simply exhibit a fold. The branching point itself can in fact only be observed experimentally if we systematically vary an imperfection parameter as well as the loading parameter. This branching point is indeed a manifestation of the cusp catastrophe, and the necessity of a two-parameter scan confirms Thom's prediction.

This static bifurcation is nicely displayed in the *symmetry-breaking* instability of the elastic arch of Figure 16. Here, by scanning both the load P and a controlled offset f, Roorda[27] was able to determine experimentally the two-thirds power-law imperfection sensitivity in his classic experiments at University College. We see that, as Thom predicts, a two-parameter scan has succeeded in locating a cusp, remembering that it is not clear a priori that a *single* controlled imperfection would be adequate to balance out all the inherent manufacturing errors in the test arch as this one does at $f = f_0$. The experimental cusp is off-centre as we predicted,

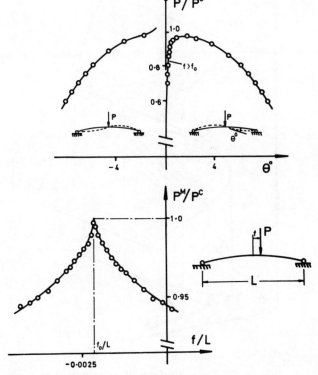

Figure 16 Experiments of Roorda on the buckling of a shallow arch[27]

but according to Thom it should also be tilted: there is a little asymmetry near the cusp point, but there is little evidence of much overall tilt. Of course Thom does not say *how much* it should be tilted!

The topologists have usually drawn three-dimensional equilibrium *surfaces*, and Figure 17 shows the equivalence of our symmetric point of bifurcation and the conventional cusp diagram of Zeeman.[10,49] Here constant ε slices retrieve the bifurcation diagram, and we notice that the stability boundary, a trace of folds, projects into the horizontal control space as our two-thirds power-law cusp.

The *stable*-symmetric point of bifurcation, as exemplified by the Euler column, is also classified by Thom as a cusp due to its essential topological similarity. We shall examine the cusp in Euler buckling in the following chapter.

As an example of such a stable cusp, we turn now from conservative mechanical systems to fluid mechanics, and Figure 94 of Chapter 7 illustrates the classical hydrodynamic instability of Couette flow between rotating cylinders. If the cylinders are long, so that the end effects can be ignored, the primary circular flow loses its stability at a stable-symmetric, super-critical bifurcation as the angular velocity is increased. This essentially static bifurcation gives rise to steady

16

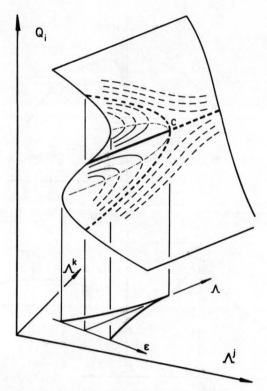

Figure 17 The equivalence of an unstable cusp
and an unstable-symmetric point of bifurcation

cellular vortices named after G. I. Taylor, as illustrated in the top diagram of Figure 94.

This phenomenon has very recently been investigated experimentally by Benjamin,[50] using, however, a cylinder of very finite length, as shown in the lower diagram. Here the end conditions destroy the simple primary flow, and Benjamin treated the angular velocity and the variable length as two independent parameters controlling the transition between two- and four-cell flows. In this way he obtained the tilted cusp shown on a plot of length L against the Reynolds number R, and his results are drawn schematically in the three dimensions of Figure 95. This cusp controlling the morphogenesis of Couette flow is essentially similar to that controlling cell differentiation as suggested by Zeeman in his contribution to developmental biology, which we shall present in Chapter 5.

The cusp, being a simple catastrophe with an interesting, but draw-able, three-dimensional structure, has attracted a lot of interest, and is unfortunately synonymous in many people's minds with catastrophe theory itself. Zeeman[10,49,51] postulated many interesting applications of it in the social and inexact sciences which, while extremely stimulating, have provoked some controversy.[52-55] One of Zeeman's very plausible applications is that in

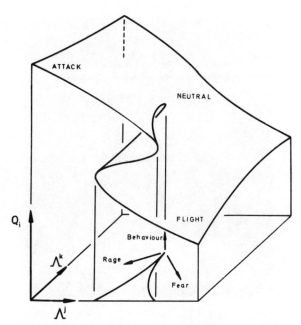

Figure 18 A cusp catastrophe illustrating the bifurcational response of an animal when subjected simultaneously to rage and fear

psychology, describing the conflicting drives of rage and fear as illustrated in
Figure 18.

The cusp, in the form of the symmetric bifurcations, does of course arise
throughout the mathematical sciences, and catastrophe theory has provided a
valuable service in demonstrating its *structural stability* under the operation of
only two unfolding control parameters.

One area in which folds and cusps arise in the presence of a well-defined total
potential energy is that of hydrostatic meniscus instabilities. Here we might
mention the work of Michael and his co-workers in electrohydrostatics[56-59]
which are important in the breakdown of electrical insulation, and the work of
Taylor on the making of holes in a sheet of fluid.[60] A review of meniscus instabilities, in which he relates these to our present work and in particular observes a
neat analogy with the dead and rigid loading of engineering structures,[47] is
given recently by Michael.[61]

Before leaving the cusp, we might note that it arises frequently in optics, and
can in fact be seen in a glass of milk standing in the sun; this is illustrated in
Photograph 1 and in the computer-drawn ray diagram of Figure 19. This has
deep connections with the theory because light follows paths of minimum (or
strictly stationary) time. In fact a ray from A to B via a curved mirror C takes up
the configuration that an elastic band would adopt if held at A and B and allowed
to run freely through a bead constrained to lie on C.

Highly acclaimed work by Berry, Nye, and others[62-67] has identified in light

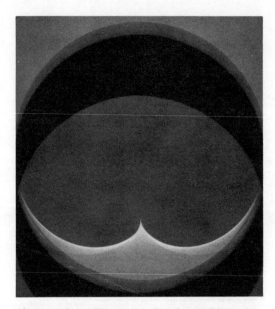

Photograph 1 The cusp to be observed in a glass of milk standing in the sun, an example of a cusp catastrophe produced as a caustic reflected from a cylindrical mirror. This picture was recorded directly on sensitive photographic paper placed under a ring of aluminium foil, made with the help of my children, Helen and Richard

caustics both the simple cusps of Photograph 2 and many of the higher-order catastrophes (Photograph 7 of Chapter 4), while Nye has used catastrophe theory to describe events in evolving three-dimensional vector fields.[68,69]

1.7 The hyperbolic umbilic and hill-top branching

We turn now to the hyperbolic umbilic, which is one of the higher-order catastrophes in Thom's list of seven. This singularity has three control parameters and two active coordinates, and so its equilibrium surfaces would have to be drawn in five-dimensional space. This is not feasible, so we consider instead the three-dimensional *control space* of Figure 20.

Here the right-hand diagram shows the *stability boundaries* in control space, corresponding to the imperfection sensitivity plot of a cusp. As the controls are varied, the equilibrium solutions vary, and they undergo a topological change of form whenever we cross one of the critical surfaces. In the diagram a solid circle denotes a stable energy *minimum*, an open circle denotes an unstable energy *maximum*, and a half-solid circle denotes an unstable *saddle-point*. The four regions of control space correspond to the indicated equilibrium solutions, there being, for example, no equilibrium solutions at all in front of the surfaces, four

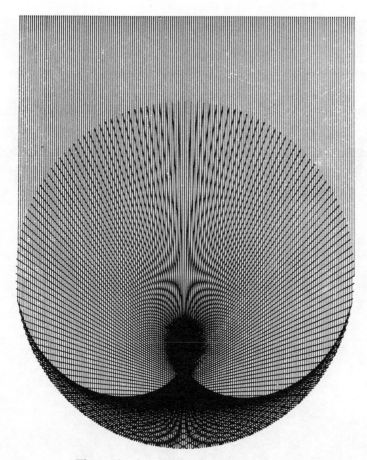

Figure 19 A computer-drawn cusp caustic

equilibrium solutions behind the surfaces, and two in the intermediate regions.

So as we move about in control space the system will suffer a stability change every time we cross one of the critical surfaces. Taking the various *routes*, A, B, and C, through the central hyperbolic umbilic point gives rise to the *path* phenomena shown on the left,[70] and it is this preoccupation with routes and paths that leads us to our sub-classification of Thom's catastrophes. The necessary sub-classifications arising from the introduction of preferred control parameters have indeed now been made by the topologists, notably by Wassermann[71-73] and more recently by powerful work of Golubitsky and Schaeffer.[74-76]

An attractive illustration of the hyperbolic umbilic arises in the interactive buckling of stiffened plates. Figure 21 shows two modes of deflection of an axially compressed plate which might have been designed to buckle simultaneously, according to a simple optimization philosophy,[77] into the two mode-forms shown. Here the load Λ and geometrical imperfections in the two deflection

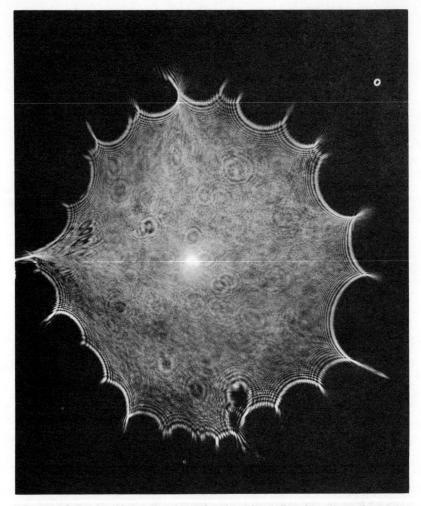

Photograph 2 Caustic produced by refraction of laser light by an irregular water-droplet 'lens' on a flat glass plate. The caustic consists of smooth *fold* curves meeting at *cusp* points, in conformity with Thom's theorem which states that these are the only stable morphologies in the plane. Reproduced with the permission of M. V. Berry

modes must be viewed as our control parameters, and the computed three-dimensional imperfection sensitivity diagrams due to Hunt[78] are shown in Figure 22.

The left-hand diagram shows the critical surfaces under simultaneous buckling, while the right-hand diagram shows them when the buckling loads have been split by a fourth control parameter. The topological *similarity* of these two forms is a consequence of the *structural stability* of the hyperbolic umbilic catastrophe; this guarantees that the form cannot be destroyed by perturbations over and

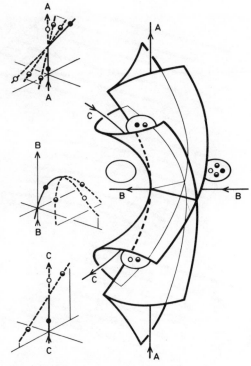

Figure 20 Control routes through a hyperbolic umbilic catastrophe

above the three active controls needed for the *universal unfolding* of the
singularity.

Rear views of these attractive surfaces are shown in Figure 23, and this work
has now been further extended by Hunt in a number of papers published by the
Royal Society.[79-81] This new work increases the range of validity of the
asymptotic analysis by embedding the hyperbolic umbilic in a controlling
parabolic umbilic catastrophe.

A second illustration of the hyperbolic umbilic in a well-defined gradient
problem due to Thompson and Shorrock[82,83] arises in the bifurcational
instability of a mechanically stressed atomic lattice that we shall discuss in
Chapter 4. Here a crystal pulled in pure tension can develop a symmetry-
breaking shearing strain, and the superposition of a lateral compression can yield
a simultaneous instability at the hill-top branching point of Figure 20. The
stability boundary in the three-dimensional control space is now the *failure-stress*
locus of Figure 24.

A novel application of the *elliptic* umbilic catastrophe has been given by Berry
and Mackley[84,85] to describe the unfolding of an unstable fluid flow with the
generation of a circulating vortex.

Photograph 6 of chapter 4 of a home-made cardboard model of the hyperbolic
umbilic stability boundaries illustrates the intricate features of this singularity,

22

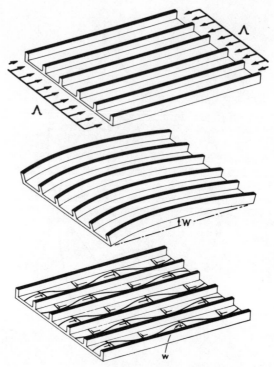

Figure 21 Two modes of deflection in the buckling of
stiffened plates

which include a cusped edge, it being common to find the low-order catastrophes
in the environments of the higher catastrophes.

1.8   Higher-order catastrophes and structural optimization

Higher-order bifurcations, falling outside the list of seven elementary catastrophes and involving a multiplicity of buckling modes are often generated by a
process of structural optimization combined with built-in symmetries.[77] These
bring with them complex secondary bifurcations[86-89] which play important
roles in planetary evolution and chemical thermodynamics, as we shall see.

The compound instability of a simultaneously buckling elastic plate has been
nicely discussed by Poston and Stewart in their excellent and comprehensive
book[11], where they identify the physical parameters necessary to unfold the
corresponding double-cusp catastrophe.

A structural example of highly complex bifurcational behaviour is provided by
a complete spherical shell under uniform external pressure as analysed by
Koiter,[90,91] and the theoretical collapse of such a shell is summarized in Figure
25. Here under rigid loading[47] we have a dynamic jump at constant volume, as
has been observed experimentally for an imperfect shell.[92,93]

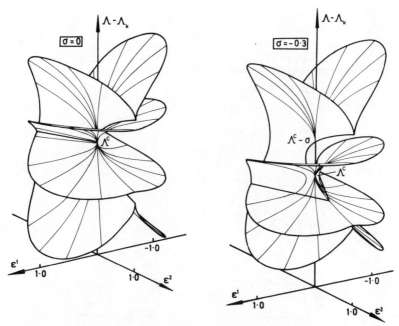

Figure 22 The hyperbolic umbilic catastrophe in the imperfection sensitivity of stiffened plates showing the structural stability of the catastrophe boundary: front views, due to Hunt[78]

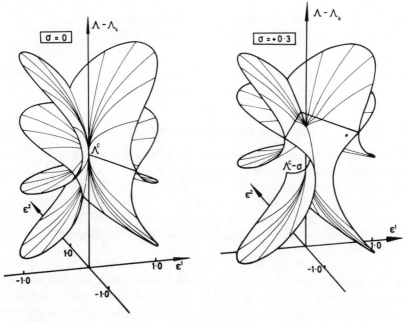

Figure 23 Rear views of the hyperbolic umbilic catastrophe due to Hunt[78]

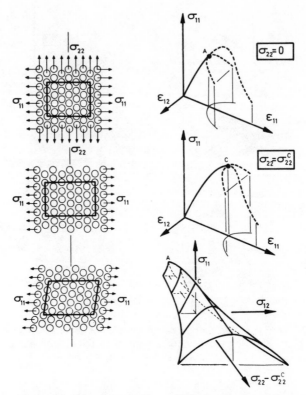

Figure 24 The hyperbolic umbilic in the symmetry-breaking instability of a mechanically stressed atomic lattice

The classification of elementary catastrophes of co-dimension less than or equal to five has been presented by Zeeman[94] in 1976.

1.9 Galloping and flutter of suspension bridges

We shall leave the static bifurcations now and turn instead to dynamic bifurcations. These are typified by the emergence not of a statical equilibrium path but of a trace of limit cycles.

Now when a steady wind blows across a flexible elastic structure it can induce and maintain large amplitude oscillations such as those that destroyed the Tacoma Narrows suspension bridge, shown just prior to collapse in Photograph 11 of Chapter 9.

Three distinct mechanisms of aeroelastic excitation are unimodal *galloping*, *vortex resonance*, and bimodal *flutter*, and it now seems the Tacoma bridge was destroyed by a combination of more than one of these phenomena. We shall, however, for the moment focus attention on simple galloping, which is observed in its purest form in the dangerous vibrations of ice-coated power cables.

Consider, for example, the wind of velocity V blowing past the square prism of

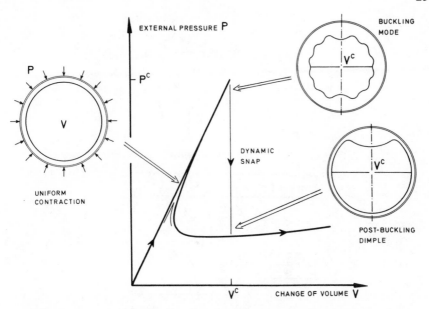

Figure 25 A constant-volume snap during the bifurcational collapse of a complete spherical shell

Figure 26, which is constrained by a spring and a dash-pot to move vertically at right angles to the wind. Now when the prism moves downwards with velocity $\dot{y}$ the velocity of the air *relative* to the body is given by V_R in the triangle of velocities drawn. This relative wind will give rise to a vertical component of force as shown. Under a quasi-static assumption the force coefficient C is simply dependent on the angle α which in turn depends on $\dot{y}$. Two typical variations of C with α are shown in the lower diagram. These were determined by Parkinson, and Brooks[95,96] from aerodynamic tests on stationary inclined sections.

We see that the wind is essentially providing *negative damping*, giving an instability at a dynamic Hopf bifurcation at a wind speed proportional to the dash-pot constant r. Notice that if we had no structural damping the system would be 'destabilized' by an *infinitesimal wind*, which emphasizes the *structural instability* of a completely conservative system.

Different cross-sections give different $C(\alpha)$ characteristics and hence different galloping phenomena, and some forms outlined by Novak[97] are discussed in detail in Chapter 9. Both stable and unstable bifurcations are possible, together with hysteresis jumps from *dynamic folds* at the coalescence of stable and unstable limit cycles.

A useful modern summary of some of these flow-induced oscillations is contained in the recent book of Blevins[98].

1.10 Emergence of order in biochemical reactions

Bifurcations are key factors in the spontaneous emergence of ordered patterns in chemical and biochemical systems, as highlighted by the work of the Brussels

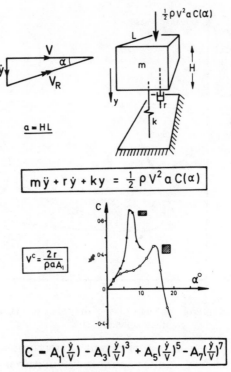

$$m\ddot{y} + r\dot{y} + ky = \tfrac{1}{2}\rho V^2 a C(\alpha)$$

$$V^c = \frac{2r}{\rho a A_1}$$

$$C = A_1\left(\frac{\dot{y}}{V}\right) - A_3\left(\frac{\dot{y}}{V}\right)^3 + A_5\left(\frac{\dot{y}}{V}\right)^5 - A_7\left(\frac{\dot{y}}{V}\right)^7$$

Figure 26 The galloping of a square prism in a steady wind: model and aerodynamic characteristic due to Parkinson and Brooks.[95] Reproduced with the permission of the Am. Soc. of Mech. Engrs.

group of I. Prigogine[99,100]. This spontaneous creation of spatial and temporal organization is vital for the understanding of morphogenesis in developmental biology, and is discussed in Chapter 5.

This creation of order can, by reason of the *second law of thermodynamics*, only occur in an *open* system, which moreover must be essentially non-linear in behaviour. For such a system a self-organizing process is accompanied by an instability of a path of steady states, and beyond this instability an initially homogeneous system has the possibility of attaining an ordered configuration or *dissipative structure*.

A physical example of this is provided by the well-known Zhabotinski reaction, and spiral chemical waves that can emerge from a shaken homogeneous chemical mixture are shown in Figure 74 of Chapter 5 traced from a photograph due to Winfree.[101] These can be compared with the dramatic shapes of living organisms reproduced in D'Arcy Thompson's classic book, *On Growth and Form*.[102]

The reaction kinetics of the Zhabotinski phenomenon are rather complex, and much theoretical interest has been focused on a trimolecular model system, the so-called Brusselator.[100] With no spatial distribution this system exhibits a *dynamic bifurcation* in the phase space of two chemical concentrations, x and y, and the growing limit cycles are shown in Figure 78 of Chapter 5. Here the imposed concentration of a third chemical, B, is taken as a control parameter, and the stable limit cycles for B greater than its critical value represent maintained oscillations in the chemical composition of the mixture. It is felt that oscillations of this type could account for the mysterious biological clocks.

If this model reaction is distributed in space, so that chemical diffusion can take place, it gives rise to spontaneous patterns as with the Zhabotinski reaction, and it is speculated that spontaneous organization of this type might account for the emergence of life from the primeval soup.

Similar, though not identical, oscillations are predicted by the Lotka–Volterra evolution equations derived in Chapter 6 for the population dynamics of a prey–predator ecology.[103] Some attempts have been made to employ catastrophe theory in evolution,[104,105] and the role of bifurcations and dynamic complexity in ecological systems is discussed by May.[106]

1.11   Non-conservative problems of fluid loading

We have already discussed galloping at some length, and we shall here pass quickly over vortex resonance, simply noting that the *fluid oscillator* models of

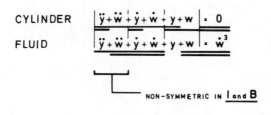

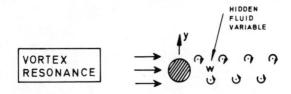

CYLINDER $\quad |\ddot{y}+\ddot{w}+\dot{y}+\dot{w}+y+w| = 0$

FLUID $\quad |\ddot{y}+\ddot{w}+\dot{y}+\dot{w}+y+w| = \dot{w}^3$

NON-SYMMETRIC IN $\mathbf{I}$ and $\mathbf{B}$

IWAN and BLEVINS : TERMS INDICATED BY UPPER LINE

HARTLEN and CURRIE : TERMS INDICATED BY LOWER LINE

Figure 27   Fluid oscillator equations for vortex resonance

28

Iwan and Blevins[107] and Hartlen and Currie[108] generate the equations of two coupled non-linear oscillators, with a possibly non-symmetric inertial matrix, as shown in Figure 27. Hopf bifurcations in wind-induced vibrations are discussed by Poore and Al-Rawi.[109]

We examine now three well-defined problems of fluid loading to see the types of generalized force that can arise and to make some preliminary mechanics classifications.

We start with the work of Dowell[110,111] on classical aerodynamic flutter,[112] which is summarized in Chapter 9. Here a supersonic wind stream blows past an elastic panel generating a stable dynamic bifurcation, presumably of the Hopf type, with a growing stable limit cycle. The matrices and non-linear terms from a two-mode harmonic analysis are summarized, and we observe that the fluid flow gives us positive definite *damping* proportional to the velocity U and an antimetric circulatory matrix proportional to U^2.

We secondly consider the simply supported pipe carrying a flowing fluid analysed by Holmes.[113] The matrices and non-linearities for his two-mode harmonic analysis are given in Chapter 9. Here we see the flow is giving an antimetric *gyroscopic* matrix associated with coriolis forces proportional to the fluid velocity U and a symmetric conservative matrix associated with centrifugal

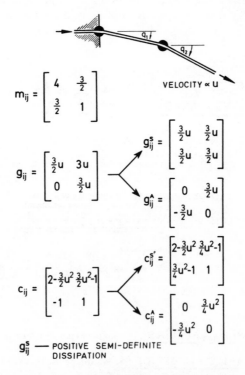

Figure 28 Force matrices from Benjamin's analysis
of two articulated pipes carrying a fluid flow

forces proportional to U^2. This latter is identical in form to the compression matrix of an applied axial compressive load. Since the coriolis forces, being antimetric, do no work, and since complete structural damping is assumed to operate, this system cannot exhibit a limit cycle and therefore cannot yield a Hopf bifurcation: in fact it exhibits *divergence* entirely analogous to the buckling and post-buckling of a compressed Euler column. It is perhaps worth noting that at a load *above the second Euler load*, motion away from the trivial unstable state can, however, exhibit growing oscillatory behaviour akin to flutter due to the coriolis terms: as the system begins to 'roll' off the energy hill, the coriolis force at right angles to the motion curves the path sufficiently to give oscillatory behaviour.

We finally recall the *linear* flutter studies of an articulated cantilevered hosepipe due to Benjamin.[114] The matrices of Figure 28 show that the fluid forces can be decomposed into positive semi-definite *dissipation* and workless *gyroscopic* forces both proportional to U, and symmetric *conservative* and antimetric *circulatory* forces both proportional to U^2. The flow-induced centrifugal forces c_{ij} proportional to U^2 are in fact *statically equivalent* to a follower force at the discharging tip.

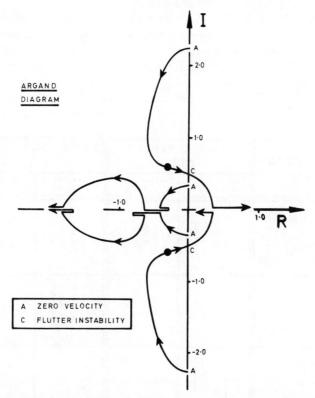

Figure 29 Movement of the roots of the characteristic equation in the complex plane for Benjamin's articulated pipes

The movement of the roots of the characteristic equation are shown in the complex plane of Figure 29, and because Benjamin includes no structural damping the roots for zero fluid velocity lie on the imaginary axis at A. A small fluid flow gives us negative real parts, but a *dynamic instability* finally sets in at C where two complex conjugate roots cross the imaginary axis. In the absence of a complete non-linear formulation of the fluid and structural forces we can simply speculate that this dynamic instability at C is a Hopf bifurcation of a stable or unstable type. Some relevant non-linear studies can perhaps be mentioned here.[115-117]

An analogous study of a continuous cantilevered pipe has been made in a series of papers by Paidoussis, to be discussed later in Chapter 9. The continuous cantilevered pipe has also been examined by Bishop and Fawzy,[118] including also a discussion of its forced oscillations; and Paidoussis and Deksnis[119] have drawn attention to a surprising paradox.

In each of the three problems discussed in detail, the fluid forces had been *linearized*, and the resulting matrices are summarized in Figure 30. It is seen that the studies of Dowell and Holmes have rather opposite contributions, while Benjamin's cantilever has forces of all types.

These examples of real fluid loading lead us to suggest the classification of forces of Figure 31, the term *impressive* being suggested for rate-dependent forces that do *positive work on the system*, as do the aerodynamic forces in galloping. The resulting classification of systems follows in Figure 32. Notice that all systems are assumed to have *inertial, elastic,* and *positive definite dissipative forces,* the latter being essential for good modelling of the real world, as we shall emphasize in the following section. With this assumption about dissipation, a minimum of the

FLOW-INDUCED FORCES	RATE-DEPENDENT $g_{ij}\dot{q}_j$		DISPLACEMENT-DEPNT. $c_{ij}q_j$	
	SYMMETRIC DISSIPATIVE	ANTIMETRIC GYROSCOPIC	SYMMETRIC ELASTIC	ANTIMETRIC CIRCULATORY
BENJAMIN 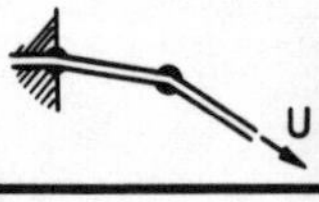	$U\begin{bmatrix} 1 & 1 \\ 1 & 1 \end{bmatrix}$ POS. S-DEF.	$U\begin{bmatrix} 0 & 1 \\ -1 & 0 \end{bmatrix}$	$U^2\begin{bmatrix} -2 & 1 \\ 1 & 0 \end{bmatrix}$ INDEF.	$U^2\begin{bmatrix} 0 & 1 \\ -1 & 0 \end{bmatrix}$
HOLMES 		$U\begin{bmatrix} 0 & -1 \\ 1 & 0 \end{bmatrix}$	$U^2\begin{bmatrix} -1 & 0 \\ 0 & -4 \end{bmatrix}$ NEG. DEF.	
DOWELL 	$U\begin{bmatrix} 1 & 0 \\ 0 & 1 \end{bmatrix}$ POS. DEF.			$U^2\begin{bmatrix} 0 & -1 \\ 1 & 0 \end{bmatrix}$

Figure 30   Summary of the flow-induced forces from three analyses

CLASSIFICATION OF FORCES

●	INERTIAL	$T_{ij}(0)$ SYM. P.D.	$m_{ij}\ddot{q}_j$
●	DISSIPATIVE	$D_{ij}(0)$ SYM. P.D.	
	GYROSCOPIC	G_{ij} ANTIMETRIC	$g_{ij}\dot{q}_j$
	IMPRESSIVE	I_{ij}	
●	ELASTIC	$U_{ij}(0)$ SYM. P.D.	$c_{ij}q_j$
	CONSERVATIVE	$W_{ij}(0)$ SYM. $\rbrace V_{ij}(0)$	
	CIRCULATORY	C_{ij}	
	NONLINEARITIES		$N_i(q_j,\dot{q}_k,\ddot{q}_l)$
●	Assumed to be always present		$= 0$

Figure 31 A suggested classification of mechanical forces

CLASSIFICATION OF SYSTEMS

FORCES / SYSTEM	RATE DEPENDENT		DISPLACEMENT DEPENDENT		EXAMPLES	ENERGY THEOREM $V_{MIN} \rightleftharpoons$ ST.
	GYROSCOPIC (Workless)	IMPRESSIVE (Working)	CONSERVATIVE (V-derivable)	CIRCULATORY (Not V-derv.)		
CONSERVATIVE			●		BUCKLING OF AN EULER STRUT	YES
GYROSCOPIC	●		●		DIVERGENCE OF A PIN-ENDED PIPE	YES
IMPRESSIVE		●	●		GALLOPING OF A BLUFF STRUCTURE	NO
CIRCULATORY			●	●	FLUTTER OF AN AIRCRAFT PANEL	NO
GYRO. CIRCUL.	●		●	●	FLUTTER OF A CANTILEVERED PIPE	NO

ALL SYSTEMS ARE ASSUMED TO HAVE INERTIAL, DISSIPATIVE AND ELASTIC FORCES

Figure 32 A suggested classification of systems with examples of each and the range of validity of the energy theorem

32

total potential energy of a discrete system becomes both *necessary and sufficient* for the stability of conservative *and* gyroscopic systems as indicated.[120]

Topical and technically very important problems of flow-induced instabilities are the Mathieu instabilities experienced by models of the proposed *tethered* oil platforms (tension-leg platforms, tethered-buoyant platforms, articulated towers, etc.) under the action of a steady train of surface waves, as dramatically predicted by Rainey.[121–123]

1.12 Dynamical systems theory and topological stability

We believe that an important programme of engineering research should aim to apply the relatively new concepts and ideas of dynamical systems theory to these and other flow-induced instabilities as stressed by Holmes, Marsden, Rand, and Chillingworth in a number of significant papers.[113,124–134]

This body of theory, due to the topological school of our introduction, can be seen in the Proceedings of a New York Conference on Bifurcation Theory,[135] in the book of Abraham and Marsden,[16] and in related works.[136–142] It embraces the key role of *structural stability*, which demands that our modelling should be *non-linear* (at a bifurcation point) and should *always include damping*. The inclusion of the latter serves to eliminate such well-known engineering paradoxes as the finite destabilization due to infinitesimal damping[143,144] shown in Figure 33.

In the wider field of dynamical systems theory, no complete classification corresponding to that of elementary catastrophe theory is yet available, but it is already established that the Hopf bifurcation and the saddle-node (or fold

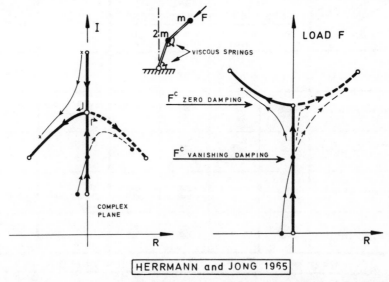

Figure 33 An example of a structurally unstable model in which infinitesimal damping generates a finite destabilization of a column carrying a follower force

catastrophe) are the only structurally stable *local* bifurcations observable under the operation of a single control parameter. So a positive benefit of showing that a fluttering cantilevered hosepipe exhibits a Hopf bifurcation is that its structural stability and experimental observability under one control would be guaranteed.

One powerful concept of dynamical systems theory emphasized by Chillingworth and Holmes is that of a centre manifold on which the simple Hopf bifurcation of a two-dimensional phase space can be found embedded in a phase space of higher dimension[45].

1.13 Chaos, turbulence, and strange attractors

One of the most exciting products of the foregoing dynamics research is the new concept of a strange attractor. This emerged from the work of Lorenz on atmospheric effects,[145] and the implied generation of chaos in the responses of a simple deterministic differential equation was seized upon by Ruelle and Takens[146] as a possible explanation of hydrodynamic turbulence. This latter field has still a number of unexplained features as we shall indicate in Chapter 7.

The essentially chaotic phase portraits of these *strange attractors* have been observed in the dynamical response of *very* simple non-linear dynamical systems on a three-dimensional phase space. Indeed, Rossler has said:[147,148] 'If oscillation is *the* typical behaviour of two-dimensional dynamical systems, then chaos, in the same way, characterizes three-dimensional continuous systems.' Here chaos is 'an infinite number of unstable periodic trajectories and an uncountable number of non-periodic recurrent ones'.

Such chaotic behaviour has now been observed by Holmes and co-workers in quite simple mechanical systems,[149-156] notably in the laterally forced vibrations of a slightly buckled beam. This could have a significant impact on the interpretation and validity of numerical time integrations and averaging techniques. A book on this expanding field should be noted[157].

As a brief introduction to this field we give in Chapter 8 a simple computer study of the strange attractor proposed by Hénon[158] which is essentially a difference equation designed to model the Poincaré map of a continuous system. Similar difference dynamical systems have been studied computationally by Hsu and co-workers.[159-161]

1.14 Concluding remarks

I hope that I shall in this book succeed in conveying some of the delights of instabilities, bifurcations, and catastrophes, both in analytical mechanics and in the wider mathematical sciences. So perhaps I can end this introductory survey with a quotation from Cornelius Lanczos:[162]

> Analytical mechanics is much more than an efficient tool for the solution of dynamical problems encountered in physics and engineering. There is hardly a branch of the mathematical sciences in which

abstract rigorous speculation and experimental evidence go together so beautifully and support each other so perfectly.

There is a tremendous treasure of philosophical meaning behind the great theories of Euler and Lagrange, and of Hamilton and Jacobi ... a source of the greatest intellectual enjoyment to every mathematically minded person.

CHAPTER 2

The Buckling of Engineering Structures

Historically and logically, the buckling of elastic structures lies at the heart of stability theory, and its simple physical models form a valuable introduction to our more esoteric applications. For this reason we shall now look at the response and simple analysis of some buckling problems, ending with a sketch of current research in this area.

2.1 Simple experiments on a flexible column

Suppose that we sharpen the ends of a flexible strip of springy metal or wood of length L and compress it axially as a column using a loaded piston as shown in Figure 34. Adding slowly the dead load P, we could measure the central side deflection Q and plot a (P, Q) load-deflection graph. As an alternative to this dead loading, we could of course use a rigid screw device[47] to impose an end-shortening; the equilibrium states would then be the same, but the regions of stability would be changed.

Proceeding with our dead loading, we would find that very little Q would be recorded until P approached a critical value P^C. Close to this load, the side deflection would be observed to grow rapidly with increasing load, as illustrated by the curve OKN of the figure, this *buckling* growth being limited by a *post-buckling* stiffening of the system. At N the strut is in the highly bent configuration familiar to us when we lean on a garden cane. Despite the rapid growth of the deflection near P^C, the *natural* equilibrium path OKN is everywhere stable and movement along it is smooth and reversible, as indicated by the arrows. We are assuming throughout that there is never any material failure or yielding, so the behaviour is *elastic* and the column always springs straight again on removal of the load.

Now it is a common experience that a cane can be held bent in either direction with either positive or negative Q, and at a high compressive load it would be found that the column could be moved by hand over to the stable *complementary* state M. If we were then to unload the column we would find that at the minimum point J the column would jump dynamically back to state K on the natural loading path, there being essentially no change in the dead load during this fast dynamic 'snap-through'. Subsequent loading and unloading would now simply induce the previously observed smooth natural behaviour. The jump at the *limit point* J is an example of a *fold catastrophe*.

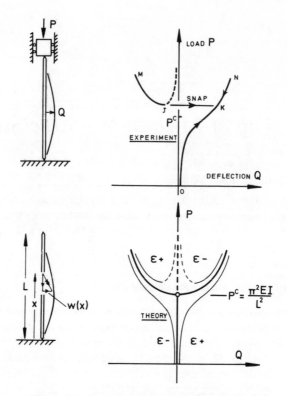

Figure 34 Experimental and theoretical behaviour of an axially compressed Euler column

Suppose we now set out to model this behaviour mathematically. We might first try to use simple engineering *bending* theory in which we approximate the curvature of a short element of beam as $\partial^2 w/\partial x^2$, where $w(x,t)$ is the lateral deflection of a cross-section distance x from the lower support at time t. Setting this curvature proportional to the bending moment evaluated in the *undeflected* state, which is everywhere zero for a perfect strut, we would just find the simple *fundamental solution* $w = 0$. This is, of course, a true equilibrium state for a perfectly straight column, but tells us nothing about the expected buckling.

Now just as the tensioning of a guitar string raises the natural frequency of the straight configuration, so the compressing of a straight elastic column lowers the natural frequency, until it finally becomes zero at a critical load P^C. Indeed, a linear vibration analysis, containing quadratic energy terms, shows that the square of the fundamental natural frequency of the first harmonic $w = Q \sin \pi x/L$ drops linearly to zero at the Euler buckling load $P^C = \pi^2 \, EI/L^2$. Here EI is the bending stiffness, being the product of Young's modulus of elasticity of the material, E, and the second moment of area of the column cross-section, I.

This critical load, which is of fundamental importance in structural engineering, can be obtained purely by statics in an engineering *buckling* analysis if we again approximate the curvature as $\partial^2 w/\partial x^2$ but write the bending moment in a *deflected* state as Pw. The result of such a linear *eigenvalue* analysis predicts that the deflection will tend to infinity at a set of critical loads corresponding to the harmonic deflections

$$w = Q_n \sin \frac{n\pi x}{L}$$

where $n = 1, 2, 3, \ldots$, the lowest critical load being P^C corresponding to $n = 1$.

A complete non-linear post-buckling static analysis using the more complex exact curvature expression gives the complete equilibrium picture for a perfect column which was first determined by Euler in his classic work[2]. It is shown by the heavy curves in Figure 34 for values of the load around $P = P^C$, a solid curve denoting a stable equilibrium path and a broken curve denoting an unstable equilibrium path.

To reconcile this picture with our experiments we must finally put initial imperfections into our mathematical modelling, such as an initial out-of-straightness of the column. Such imperfections, in the first harmonic for example, destroy the trivial fundamental solution and round off the bifurcation at P^C, as shown in the complete bifurcation diagram. Here the light curves represent the equilibrium paths of two imperfect systems, one with a positive value of an

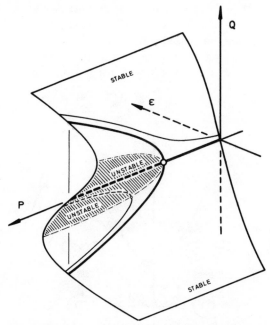

Figure 35 Three-dimensional representation of the equilibrium paths of an Euler column showing the displacement Q as a function of the load P and an imperfection ε

38

imperfection parameter ε and the other with a negative value of ε. Once again solid/broken curves represent stable/unstable paths respectively. This stable-symmetric branching behaviour has been studied extensively in general terms by Thompson and Hunt[36].

We can finally draw a three-dimensional *equilibrium surface* giving the generalized coordinate, or active state variable Q, in terms of the two control variables P and ε as shown in Figure 35. This identifies the stable-symmetric point of bifurcation of our Euler strut as a stable *cusp catastrophe*. Discussions of Euler buckling in terms of catastrophe theory have been given by Chillingworth[163] and Zeeman.[164]

Before beginning our analysis of a continuous strut, we now look at a one-degree-of-freedom link model to introduce some of the basic ideas. This elastically restrained link can be thought of as a simple model of a cantilevered column built in at one end, which behaves like half of a simply supported pin-ended strut.

2.2 Analysis of a cantilevered link model

Consider the light rigid link of Figure 36 with length L supported by a rotational elastic spring of stiffness k. The rod carries a concentrated mass m at its tip and a uniform gravitational field of strength g means that the cantilever is effectively loaded by a force of magnitude $P = mg$. We shall think of P as a controlled loading parameter and shall be interested in the loss of stability of the system as P is gradually increased from zero.

Taking the rotation from the vertical as our single generalized coordinate Q_1 we make first a *static analysis*. The strain energy stored in the spring is simply

$$U = \tfrac{1}{2}kQ_1^2$$

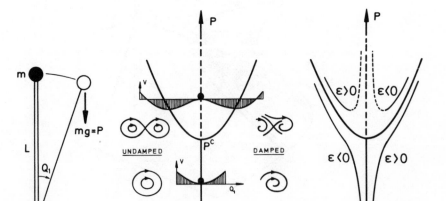

Figure 36 Static and dynamic response of a hinged cantilever model

and the fall of mass m is

$$\mathscr{E} = L(1 - \cos Q_1).$$

The total potential energy of the system is then

$$V = U - P\mathscr{E} = \tfrac{1}{2}kQ_1^2 - PL(1 - \cos Q_1).$$

The equilibrium condition is

$$V_1 = \frac{\partial V}{\partial Q_1} = kQ_1 - PL \sin Q_1 = 0,$$

giving the trivial solution $Q_1 = 0$ for all P and

$$P = \frac{kQ_1}{L \sin Q_1} = \frac{k}{L}\left(1 + \frac{1}{6}Q_1^2 + \ldots\right).$$

There are thus two equilibrium paths denoted by the heavy curves in Figure 36 and the stability of these equilibrium states is determined by forming

$$V_{11} = \frac{\partial^2 V}{\partial Q_1^2} = k - PL \cos Q_1.$$

On the trivial fundamental path, for example, the *stability coefficient* is

$$V_{11}^F = k - PL$$

and we have stability for P less than P^C and instability for P greater than P^C where

$$P^C = \frac{k}{L}.$$

In a similar manner, the post-critical equilibrium path is found to be everywhere stable, and it is clear that the total potential energy $V(Q_1)$ has a single minimum for a given $P < P^C$ and two minima separated by a maximum for a given $P > P^C$, as shown schematically in the figure.

The large non-linear dynamical motions of the system are easily discussed on the basis of these two total potential energy curves. Without damping, a centre about the trivial state for $P < P^C$ will transform into two centres separated by a saddle for $P > P^C$ as indicated. With some positive damping, a stable focus at the trivial state for $P < P^C$ will transform into two stable foci separated by a saddle as shown for $P > P^C$.

We shall not analyse these non-linear motions, but make instead a *linear vibration analysis* about the trivial equilibrium state $Q_1 = 0$. Assuming that the link itself is light, the kinetic energy is simply due to the concentrated mass m and is given by

$$T = \tfrac{1}{2}mL^2 \dot{Q}_1^2.$$

Comparing this with the standard form

$$T = \tfrac{1}{2}T_{11}\dot{Q}_1^2$$

we have

$$T_{11} = mL^2.$$

The circular frequency of small vibrations, ω, is then given from simple oscillator theory[37] as

$$\omega^2 = \frac{V_{11}}{T_{11}} = \frac{k - PL}{mL^2},$$

which drops to zero as P increases to P^C.

We observe that we have three distinct ways of determining the critical load P^C by locating:

(1) the branching of the trivial *equilibrium* path
(2) the loss of a potential *energy* minimum
(3) the vanishing of the dynamical *frequency* of vibration.

The equivalence of (1) and (2) is guaranteed by a basic theorem of elastic stability[42,43] that we have mentioned in Chapter 1. The equivalence of (2) and (3) is an established result for a general conservative mechanical system in the presence of a little damping.[37] Equivalent *linear* criteria are discussed in detail by Ziegler for a wide class of mechanical systems.[20]

We finally examine the role of an initial geometric imperfection in modifying the statical *equilibrium* paths of the system. Suppose that due to a small manufacturing error the spring is unstrained, not when the link is vertical but when it has a small initial angle $Q_1^0 = \varepsilon$. Within an arbitrary constant, the total potential energy is now

$$V = \tfrac{1}{2}k(Q_1 - \varepsilon)^2 - PL(1 - \cos Q_1),$$

so for equilibrium we now have

$$V_1 = k(Q_1 - \varepsilon) - PL \sin Q_1 = 0$$

giving

$$P = \frac{k(Q_1 - \varepsilon)}{L \sin Q_1}.$$

The trivial equilibrium path is thus destroyed, and we have now a family of equilibrium curves $P(Q_1)$ corresponding to different values of ε that round off the bifurcation of the perfect system as shown in Figure 36.

More details of the statical analysis of this hinged cantilever can be found in our earlier monograph.[36]

2.3 Large deflection beam formulation

To analyse our original continuous strut we require a non-linear large-deflection formulation, which we make here in energy terms.[36] This will allow us to use the

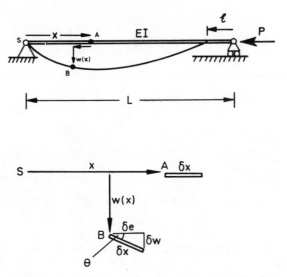

Figure 37 Deflection of a beam showing the definition of $w(x)$

Rayleigh–Ritz method for a one-degree-of-freedom analysis before leading on to a multi-mode harmonic analysis. We make then an *exact* large deflection energy formulation for the deformation of an elastic beam or column.

Consider the strut of Figure 37 of length L, simply supported to fix our ideas, and loaded by the axial force P which retains its magnitude and direction as the strut deflects. The strut is assumed to be axially rigid (inextensional) and the relevant bending stiffness is denoted by EI. Point A of the strut originally distance x from the left-hand support is displaced to B, and this displacement is resolved into an unspecified horizontal component and a vertical component w as shown. The centre-line being inextensional, the arc length SB is equal to x, and the deflected form of the strut is totally specified by the mathematical function $w(x)$ where x ranges from 0 to L. Notice that the graph of $w(x)$ does not have precisely the shape of the deflected beam because of the unspecified horizontal displacement.

The curvature χ is by definition the rate of change of angle with arc length, so

$$\chi = \frac{d\theta}{dx} = \frac{d}{dx}\sin^{-1} w' = w''(1 - w'^2)^{-1/2}$$

where a prime denotes differentiation with respect to x. The strain energy stored in an element is

$$\delta U = \tfrac{1}{2}M\chi\delta x$$

where M is the bending moment given by

$$M = EI\chi$$

so

$$\delta U = \tfrac{1}{2} E I \chi^2 \delta x.$$

The total strain energy is thus

$$U = \tfrac{1}{2} E I \int_0^L \chi^2 \, dx$$

$$= \tfrac{1}{2} E I \int_0^L w''^2 (1 - w'^2)^{-1} \, dx$$

$$= \tfrac{1}{2} E I \int_0^L (\underline{w''^2} + w''^2 w'^2 + w''^2 w'^4 + \ldots) \, dx.$$

Here the familiar underlined term is all that need be retained for a small-deflection *linear* analysis.

From the figure we have

$$\delta e = (\delta x^2 - \delta w^2)^{1/2}$$
$$= \delta x (1 - w'^2)^{1/2}$$

so the end-shortening of the column is

$$\mathscr{E} = L - \int_0^L (1 - w'^2)^{1/2} \, dx$$

$$= \int_0^L (\underline{\tfrac{1}{2} w'^2} + \tfrac{1}{8} w'^4 + \tfrac{1}{16} w'^6 + \ldots) \, dx.$$

Here the leading underlined term is again all that is needed for a linear buckling or vibration analysis.

Having made a precise static formulation we shall approximate somewhat in writing only the *linear form* of the kinetic energy. If the beam has the mass per unit length m, the kinetic energy of an element is approximately

$$\delta T = \tfrac{1}{2} m \delta x \dot{w}^2$$

where a dot denotes differentiation with respect to the time t.

Then since the total potential energy V is simply $U - P\mathscr{E}$, we have the familiar energies necessary for linear analysis:

$$V = \tfrac{1}{2} E I \int_0^L w''^2 \, dx - \tfrac{1}{2} P \int_0^L w'^2 \, dx,$$

$$T = \tfrac{1}{2} m \int_0^L \dot{w}^2 \, dx.$$

2.4 Buckling and post-buckling in the first harmonic

We make first a linear vibration and buckling analysis of a pin-ended beam or column subjected to an end compression P. The energies are thus given by the above equations and the essence of the one-degree-of-freedom Rayleigh–Ritz method is to assume a deflected mode shape with arbitrary amplitude Q. The mode shape should satisfy the geometric boundary conditions of the problem, and we here take a simple half sine wave so that

$$w = Q(t) \sin \frac{\pi x}{L}.$$

Differentiating, we have

$$\dot{w} = \dot{Q} \sin \frac{\pi x}{L},$$

$$w' = Q\left(\frac{\pi}{L}\right) \cos \frac{\pi x}{L},$$

$$w'' = -Q\left(\frac{\pi}{L}\right)^2 \sin \frac{\pi x}{L},$$

and using the results

$$\int_0^L \sin^2 \frac{\pi x}{L} \, dx = \int_0^L \cos^2 \frac{\pi x}{L} \, dx = \tfrac{1}{2}L$$

we can substitute into the linear energies to obtain

$$V = \frac{1}{2}\left\{ EI\left(\frac{\pi}{L}\right)^4 \frac{L}{2} - P\left(\frac{\pi}{L}\right)^2 \frac{L}{2} \right\} Q^2$$

$$T = \frac{1}{2} m \frac{L}{2} \dot{Q}^2.$$

Now we expect for a linear analysis to have[37]

$$V = \tfrac{1}{2} V_{11} Q^2,$$

$$T = \tfrac{1}{2} T_{11}(0) \dot{Q}^2$$

and comparing with our forms gives us our energy coefficients

$$V_{11} = \left(\frac{\pi}{L}\right)^2 \left(\frac{L}{2}\right) \left\{ EI\left(\frac{\pi}{L}\right)^2 - P \right\},$$

$$T_{11}(0) = m\left(\frac{L}{2}\right).$$

The circular frequency is then given by

$$\omega^2 = \frac{V_{11}}{T_{11}(0)}$$

so we have finally

$$\omega^2 = \frac{(\pi/L)^2 \{ EI(\pi/L)^2 - P \}}{m}.$$

We see that the square of the circular frequency decreases linearly with the compression P until it vanishes at the buckling load P^C given by

$$P^C = \left(\frac{\pi}{L} \right)^2 EI.$$

These expressions for ω^2 and P^C are in fact *exactly correct* because the assumed half sine wave is indeed the fundamental vibration mode and the buckling mode.

If we had assumed a parabolic form for $w(x)$ and proceeded as above we would have obtained the result

$$\omega^2 = \frac{(10/L^2) \{ EI(12/L^2) - P \}}{m}.$$

This represents an approximate solution and is depicted in Figure 38.

Notice that the approximate frequency and corresponding critical load are both *too high*, which is a guaranteed feature of the Rayleigh–Ritz procedure. We see also that the approximate procedure has given the correct *form* of solution and that even such a gross approximation for $w(x)$ gives a reasonably accurate answer for the circular frequency and the critical load.

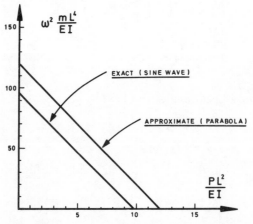

Figure 38 Frequency versus load curves for a pin-ended column showing exact and approximate results

We have so far made only linear eigenvalue analyses which for buckling simply predict neutral equilibrium at the critical load. To look at the post-buckling at this linear eigenvalue, higher non-linear terms must be retained in the total potential energy expression.

We shall here make an initial post-buckling analysis for the statical equilibrium of a compressed pin-ended Euler column using the same assumed form as before:

$$w = Q \sin \frac{\pi x}{L}.$$

The total potential energy will be $V = U - P\mathscr{E}$ where, from our formulation,

$$U = \tfrac{1}{2} EI \int_0^L (w''^2 + w''^2 w'^2 + w''^2 w'^4 + \dots) dx,$$

$$\mathscr{E} = \tfrac{1}{2} \int_0^L (w'^2 + \tfrac{1}{4} w'^4 + \tfrac{1}{8} w'^6 + \dots) dx.$$

Here the first terms of the series give the linear energy form $\tfrac{1}{2} V_{11} Q^2$ while the second terms give the non-linear energy form $\tfrac{1}{24} V_{1111} Q^4$.

Substitution and integration then gives us the energy

$$V = \frac{1}{2} EI \left(\frac{\pi}{L}\right)^4 \frac{L}{2} Q^2 + \frac{1}{2} EI \left(\frac{\pi}{L}\right)^6 \frac{L}{8} Q^4 + \dots$$

$$- P \left\{ \frac{1}{2} \left(\frac{\pi}{L}\right)^2 \frac{L}{2} Q^2 + \frac{1}{2} \left(\frac{\pi}{L}\right)^4 \frac{3L}{32} Q^4 + \dots \right\}.$$

Comparing this with a Taylor expansion we can determine the energy coefficients

$$V_{11}^{0C} = \frac{\partial^3 V}{\partial Q^2 \partial P}\bigg|^C = -\left(\frac{\pi}{L}\right)^2 \frac{L}{2},$$

$$V_{1111}^C = \frac{\partial^4 V}{\partial Q^4}\bigg|^C = \frac{3}{8} EI \left(\frac{\pi}{L}\right)^6 L,$$

where C denotes evaluation at the critical branching point at which, from previous work,

$$P = P^C = \left(\frac{\pi}{L}\right)^2 EI.$$

We can now use the result of general branching theory[37]

$$\lambda = -\frac{1}{6} \frac{V_{1111}^C}{V_{11}^{0C}} q^2$$

46

or, by direct analysis, obtain the first-order post-buckling solution

$$P - P^C = \frac{1}{8} EI \left(\frac{\pi}{L} \right)^4 Q^2 .$$

We have a stable-symmetric point of bifurcation and because this particular symmetric problem has no cubic energy coefficients at all, our one-degree-of-freedom Rayleigh–Ritz analysis has given us the correct initial curvature of the post-buckling equilibrium path.

2.5 A model column with two degrees of freedom

Before proceeding to a multi-mode analysis of our continuous column, we shall look, by way of an introduction, at an articulated model of a pin-ended strut which has just two degrees of freedom. We consider then the behaviour of the two-hinged strut shown in a *deflected* state in Figure 39.

Here three light rigid rods each of length L are hinged together to form a chain of length $3L$. One end of the chain is pivoted at a fixed point while the other is free to move axially towards this pivot. Relative rotation of the rods is resisted by two rotational springs each of stiffness k at the internal joints. These springs are unstrained when the links lie in a straight line, so that with no applied load the links lie in a straight horizontal line between the supports. The system is loaded

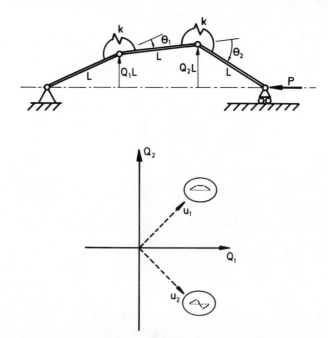

Figure 39 Deflections and modes of a two-hinged strut model

by a dead compressive load P which is assumed to retain its original magnitude and direction as the links deflect.

The system has two degrees of freedom and the vertical deflections of the internal joints are denoted by $Q_1 L$ and $Q_2 L$ as shown, so the total deflected form of the system may be fully represented by specifying the non-dimensional generalized coordinates Q_1 and Q_2. These are both equal to zero in the natural unstrained shape of the column.

The strain energy stored in the two rotational springs is

$$U = \tfrac{1}{2}k\theta_1^2 + \tfrac{1}{2}k\theta_2^2$$

which can be written as

$$U = \tfrac{1}{2}k\{\sin^{-1}Q_1 - \sin^{-1}(Q_2 - Q_1)\}^2 + \tfrac{1}{2}k\{\sin^{-1}Q_2 + \sin^{-1}(Q_2 - Q_1)\}^2$$

and expanding each of the trigonometric terms as a power series we have

$$U = \tfrac{1}{2}k(5Q_1^2 - 8Q_1 Q_2 + 5Q_2^2 + \text{higher-order terms}).$$

The end-shortening of the column due to the sideways deflection is

$$\mathscr{E} = L\,[3 - (1 - Q_1^2)^{1/2} - (1 - Q_2^2)^{1/2} - \{1 - (Q_2 - Q_1)^2\}^{1/2}]$$

and again expanding the terms as power series we have

$$\mathscr{E} = L(Q_1^2 - Q_1 Q_2 + Q_2^2 + \text{higher-order terms}).$$

Now the total potential energy of this system is simply

$$V(Q_i) = U(Q_i) - P\mathscr{E}(Q_i),$$

the latter term representing the potential energy of the applied load, so the second variation of V is simply

$$\delta^2 V = \tfrac{1}{2}k(5Q_1^2 - 8Q_1 Q_2 + 5Q_2^2) - PL(Q_1^2 - Q_1 Q_2 + Q_2^2).$$

This is precisely the quadratic form that we must inspect if we wish to examine the stability of the straight configuration, which is confirmed as an equilibrium state by the absence in the expansions of the first variation δV.

Now if we have two quadratic forms and one of them is positive definite (as is the quadratic form of the strain energy multiplying $\tfrac{1}{2}k$, since it is derived from a sum of two squares), it is always possible to find a linear change of variables to *simultaneously* diagonalize the two forms. This is in fact achieved in this particular problem by the non-singular transformation

$$u_1 = \frac{Q_1 + Q_2}{2} \quad \text{and} \quad u_2 = \frac{Q_1 - Q_2}{2}$$

which has the inverse

$$Q_1 = u_1 + u_2 \quad \text{and} \quad Q_2 = u_1 - u_2.$$

To locate the new u_i axes in Q_j space we note here that $u_1 = 0$ gives $Q_1 = -Q_2$

48

while $u_2 = 0$ gives $Q_1 = Q_2$. So the u_i axes are in this instance rectangular and at 45 degrees relative to the Q_j axes as shown.

In terms of these *principal* u_i coordinates we find by direct substitution

$$\delta^2 V = \tfrac{1}{2}k(2u_1^2 + 18u_2^2) - PL(u_1^2 + 3u_2^2)$$

and comparing this sum of squares with our standard form

$$\delta^2 V = \tfrac{1}{2}C_1 u_1^2 + \tfrac{1}{2}C_2 u_2^2$$

we find the *stability coefficients*

$$C_1 = 2k - 2PL,$$
$$C_2 = 18k - 6PL.$$

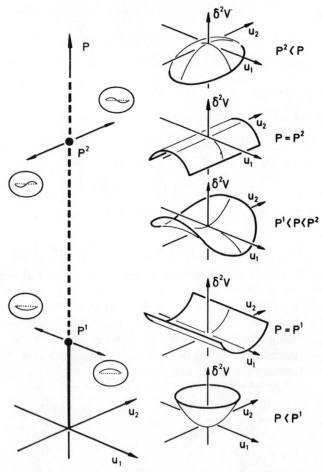

Figure 40 Three-dimensional load-deflection diagram for the linear analysis of a pinned column showing the energy surfaces at different load levels

If the applied load is zero, the straight configuration is naturally stable with

$$C_1 = 2k = \text{positive},$$
$$C_2 = 18k = \text{positive},$$

and the energy function has a local minimum as shown in the bottom right-hand diagram of Figure 40.

As the applied axial compressive load P is increased, the two stability coefficients decrease linearly with P until C_1 reaches zero at the first critical load P^1 given by

$$P^1 = \frac{k}{L}.$$

At this load the curvature of the energy surface has dropped to zero in the u_1 direction, so that the second variation yields the cylindrical form shown on the right of Figure 40. We say the column *buckles* in *mode* u_1 at this critical load.

Beyond P^1, we see that C_1 is negative while C_2 is as yet still positive. The surface of the second variation of V now has the form of a saddle-point, curving up in the u_2 direction but down in the u_1 direction as shown. The fundamental straight configuration of the model is now unstable. To supply more information, we could if wished say that it is unstable *with respect to u_1* but stable *with respect to u_2*.

As we continue to increase the applied load P, the second stability coefficient C_2 finally reaches zero at the second critical load P^2 given by

$$P^2 = \frac{3k}{L}.$$

At this second buckling load $\delta^2 V$ is again cylindrical as illustrated, and for even higher values of load it is a local maximum. We could say that our model strut buckles in mode u_2 at the second critical load P^2, although this condition could not be achieved experimentally without the addition of a constraint to inhibit the earlier buckling into mode u_1.

We emphasize that each energy surface drawn on the right-hand side of Figure 40 relates to a given fixed value of the applied load P.

The mode-forms we mentioned are easily inspected. When we speak of mode u_1 we mean the deformation that can occur with $u_2 = 0$. Now $u_2 = 0$ implies by the transformation equations that $Q_1 = Q_2$, so mode u_1 is a symmetric deformation approximating to the first harmonic of a continuous column as shown in Figure 41. Similarly, $u_1 = 0$ implies $Q_1 = -Q_2$, so mode u_2 is a skew-symmetric deformation approximating to the second harmonic of a continuous column as shown.

We notice finally that the two critical loads, P^1 and P^2, could have been determined without resort to the diagonalizing transformation by equating to zero the stability determinant

$$\left| \frac{\partial^2 V}{\partial Q_1 \partial Q_2} \right| = \begin{vmatrix} (5k - 2PL) & (-4k + PL) \\ (-4k + PL) & (5k - 2PL) \end{vmatrix} = 0$$

MODEL STRUT · CONTINUOUS STRUT

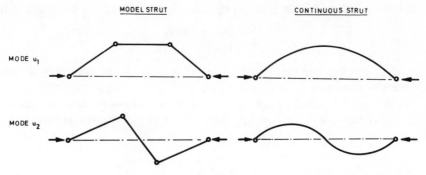

MODE u_1

MODE u_2

Figure 41 Buckling modes of the model compared with those of the real continuous strut

giving

$$9k^2 - 12kPL + 3P^2L^2 = 0.$$

This quadratic equation in PL/k can be factored as

$$\left(3 - 3\frac{PL}{k}\right)\left(3 - \frac{PL}{k}\right) = 0$$

giving

$$\frac{PL}{k} = 1 \text{ or } 3.$$

An analysis such as this that simply aims to determine critical loads by examining the second variation of V is called a *linear eigenvalue analysis*. It supplies no information about the behaviour of the system *after* initial buckling which is controlled by higher variations of V about the critical equilibrium state.

2.6 Complete harmonic analysis of a strut on a foundation

We complete this short analytical excursion by making a multi-mode linear harmonic analysis of a pin-ended strut, with or without an elastic foundation. This elastic foundation acts laterally along the length of the beam[36,37] and might be provided by a springy soil on which a horizontal strut might be lying. It can be thought of as a large number of elastic springs resisting the lateral displacement w.

The linearized energy integrals necessary for small-amplitude vibration and buckling eigenvalue analyses of elastic beams are obtained from our formulation as follows. The strain energy of bending is

$$u^B = \tfrac{1}{2}EI \int_0^L w''^2 \, dx,$$

the potential energy of a dead axial compressive load P is

$$u^P = -\tfrac{1}{2}P \int_0^L w'^2 \, dx,$$

and the strain energy of a *simple* elastic foundation supporting the beam along its length can be written as

$$u^F = \tfrac{1}{2}K \int_0^L w^2 \, dx,$$

where K is the foundation stiffness. So our total potential energy V can be written as

$$V = u^B + u^F + u^P,$$

and the kinetic energy is finally

$$T = \tfrac{1}{2}m \int_0^L \dot{w}^2 \, dx.$$

We propose to discretize the continuous elastic column in this section by employing a modal expansion for w,

$$w(x, t) = \sum_i q_i(t) M_i(x).$$

Here the amplitudes q_i are functions of time and each mode-form M_i is assumed to satisfy the *geometric* boundary conditions of the problem. Each mode-form need not satisfy the *natural* or *dynamic* boundary conditions which will be satisfied automatically as far as possible by the subsequent energy procedures.

With a finite set of modes $M_i(x)$ we are thus employing the well-known Rayleigh–Ritz energy procedure, while if we employ a complete and necessarily infinite set of functions we have a classical exact analysis, assuming the convergence of certain infinite series.

Using the dummy-suffix summation convention of Einstein we now have[†]

$$w = q_i M_i, \quad w' = q_i M_i',$$

$$w'' = q_i M_i'', \quad \dot{w} = \dot{q}_i M_i,$$

so we can write the products

$$w^2 = q_i q_j M_i M_j, \quad w'^2 = q_i q_j M_i' M_j',$$

$$w''^2 = q_i q_j M_i'' M_j'', \quad \dot{w}^2 = \dot{q}_i \dot{q}_j M_i M_j.$$

The strain energy of bending is then

$$u^B = \tfrac{1}{2}EI q_i q_j \int_0^L M_i'' M_j'' \, dx,$$

[†] with summation over any repeated subscript[36]

and comparing this with our standard linear form

$$u^{B} = \tfrac{1}{2}u^{B}_{ij}q_{i}q_{j}$$

we find the coefficients

$$u^{B}_{ij} = EI \int_{0}^{L} M''_{i} M''_{j} \, dx.$$

Similarly the potential of the load has coefficients

$$u^{P}_{ij} = -P \int_{0}^{L} M'_{i} M'_{j} \, dx$$

and the strain energy of the foundation has

$$u^{F}_{ij} = K \int_{0}^{L} M_{i} M_{j} \, dx.$$

The potential energy coefficients are then

$$V_{ij} = u^{B}_{ij} + u^{F}_{ij} + u^{P}_{ij}.$$

The kinetic energy is

$$T = \tfrac{1}{2} m \dot{q}_{i} \dot{q}_{j} \int_{0}^{L} M_{i} M_{j} \, dx$$

and comparison with

$$T = \tfrac{1}{2} T_{ij} \dot{q}_{i} \dot{q}_{j}$$

gives us finally

$$T_{ij} = m \int_{0}^{L} M_{i} M_{j} \, dx.$$

The Lagrange equations for small vibrations are now[37]

$$T_{ij} \ddot{q}_{j} + V_{ij} q_{j} = 0,$$

allowing a solution to our discretized beam problem.

As an example for a pin-ended beam we shall now take a complete Fourier expansion by writing

$$M_{i} = \sin \frac{i \pi x}{L},$$

for values of i from one to infinity. This greatly simplifies the analysis because the *orthogonality properties*

$$\left. \begin{array}{c} \displaystyle\int_0^L \sin\frac{i\pi x}{L}\sin\frac{j\pi x}{L}\,dx = 0 \\[2em] \displaystyle\int_0^L \cos\frac{i\pi x}{L}\cos\frac{j\pi x}{L}\,dx = 0 \end{array} \right\} \quad \text{for } i \neq j$$

ensure that all the energies are diagonal. That is to say, the Fourier harmonics are in fact both the buckling and normal vibration modes of the simply supported beam: the generalized coordinates q_i are already the principle coordinates u_i, and the equations of motions are already decoupled.

Then, since

$$\int_0^L \sin^2\frac{i\pi x}{L}\,dx = \int_0^L \cos^2\frac{i\pi x}{L}\,dx = \tfrac{1}{2}L$$

we have the diagonal energy coefficients

$$V_{ii} = \left\{ EI\left(\frac{i\pi}{L}\right)^4 + K - P\left(\frac{i\pi}{L}\right)^2 \right\}\tfrac{1}{2}L,$$

$$T_{ii} = \{m\}\tfrac{1}{2}L.$$

The equation for the circular frequency of the ith mode is therefore

$$\omega_i^2 = \frac{V_{ii}}{T_{ii}} = \frac{EI(i\pi/L)^4 + K - P(i\pi/L)^2}{m}.$$

This gives the normal frequencies of vibration of a simply supported elastic beam of mass per unit length m and bending stiffness EI, resting on an elastic foundation of stiffness K and carrying an axial compressive load P.

If we set ω_i equal to zero (or equivalently set V_{ii} equal to zero) we have the critical buckling loads of this system as

$$P^i = \frac{EI(i\pi/L)^4 + K}{(i\pi/L)^2}$$

and in the absence of a foundation we can set $K = 0$ to obtain the critical loads of a pin-ended column

$$P^i = EI\left(\frac{i\pi}{L}\right)^2$$

the lowest of which is

$$P^1 = \frac{\pi^2 EI}{L^2} \qquad \text{for the mode } M_1 = \sin\frac{\pi x}{L}.$$

54

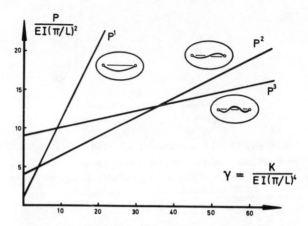

Figure 42 Critical loads of a strut on an elastic foundation against a measure of the foundation stiffness

We see that *with* a foundation of stiffness K the first harmonic ($i = 1$) does not always correspond to the lowest buckling load, as illustrated in Figure 42.

If we put the foundation stiffness to zero and the bending stiffness to zero, $K = EI = 0$, and write $P = -S$ we have the formula for the normal frequencies of a string stretched by a tension S as

$$\omega_i^2 = \frac{S(i\pi/L)^2}{m}.$$

The buckling and post-buckling of a strut on an elastic foundation with a free, un-pinned end has been discussed by E1 Naschie,[165] who has also elucidated the mechanics of ring buckling.[166]

2.7 Symmetric and non-symmetric snap-through of arches

Let us now consider some more experiments. If we take our straight flexible elastic column and tie a strong tensioned spring between the ends, we can form a bow or *shallow arch*. We shall want the arch to be able to deflect 'through' the spring, and a possible practical arrangement using two symmetrically placed springs is shown in Figure 43.

Before applying any vertical transverse loads, we can observe immediately that this pre-stressed mechanical system has three possible equilibrium states under zero load. The strut can first be placed so that it is bent either upwards or downwards, giving us the two *stable* equilibrium configurations, A and C of Figure 44. Also, if we are careful and help the balance a little, we can observe that there is an equilibrium state, albeit an *unstable* one, in which the strut is perfectly straight with the spring highly elongated; here the strut is highly compressed *above* its Euler buckling load.

If we now apply a dead transverse load P, we shall obtain the equilibrium and

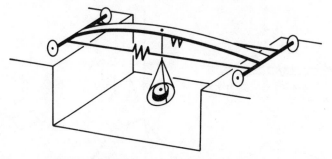

Figure 43 A physical model of a tied arch

hysteresis response sketched in Figure 44. Here two dynamic snaps bring the arch back to state A after a complete loading cycle involving negative, upwards, *P*. These snaps arise at the two *folds* or *limit points* D and E, and we can notice that the unstable dotted region of the equilibrium path between these two critical states will not be observed experimentally in a straightforward dead loading

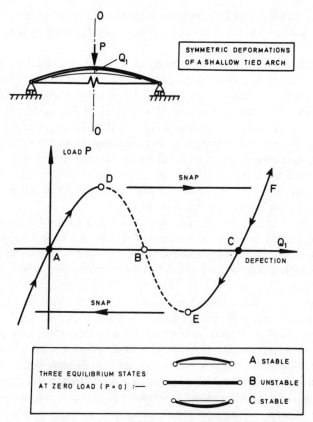

Figure 44 Symmetric deformations of a shallow tied arch, showing three equilibrium states at zero load

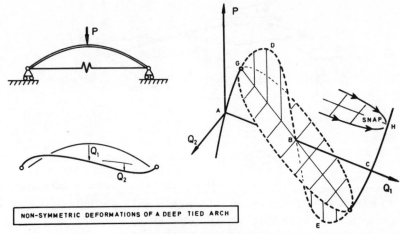

Figure 45 Non-symmetric deformations of a deep tied arch showing the equilibrium paths in three dimensions

cycle. This region can in fact be stabilized and observed by the alternative use of *rigid loading*, in which values of the deflection Q_1 are imposed by a rigid screw device.

In this simple snap-through behaviour, we can observe that on the path ADBECF the deformation of the arch is always symmetric about its centre-line O – O. If, however, we make a *deeper arch* the strut will experience second-harmonic buckling before the maximum D is reached, as illustrated in Figure 45. Non-symmetric deformations measured by Q_2 will thus develop, and the arch will exhibit the equilibrium paths shown in three dimensions in the figure.

On loading, such a deep arch would fail by a dynamic jump from G to H, the fundamental path with $Q_2 = 0$ losing its stability at the branching point G. During this jump, the arch will pass through non-symmetric states from the symmetric state G to the symmetric state H.

This symmetry-breaking instability is associated with the unstable-symmetric point of bifurcation G, which is an example of an *unstable cusp*. The symmetry-breaking must of course be triggered by a small non-symmetric disturbance or imperfection which will determine whether the dynamic route taken will involve positive or negative Q_2.

This unstable-symmetric branching implies a severe sensitivity to non-symmetric imperfections, as illustrated subsequently in Figure 49. In this figure Λ would be our load P, u_1 would be our second-harmonic amplitude Q_2, and ε would be an initial imperfection in mode Q_2. The analogous behaviour of a *fixed* arch (which can be viewed as our tied arch with an infinitely strong spring) is shown in Figure 46 where the offset of the load, f, models an initial imperfection. The lower right-hand diagram shows the imperfection sensitivity on a plot of the load-carrying capacity P^M against the magnitude of f. Experiments on such a fixed arch due to Roorda[27] have been described in connection with the earlier Figure 16, where shifting and tilting of the cusp have been discussed.

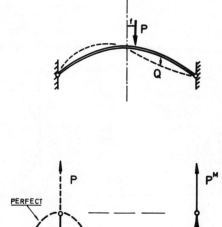

Figure 46 Load-deflection diagram and imperfection sensitivity of a deep fixed arch

2.8 Some distinct and compound branching points

The three distinct branching points that we encountered briefly in Chapter 1 are shown in detail in Figures 47 to 49, on load-deflection and imperfection sensitivity plots. They were first analysed by Koiter[17] in 1945 and have been further explored at University College London using systematic perturbation techniques;[36,167,168] this latter research has been extensively re-worked in Japan[169].

The asymmetric point of bifurcation shown in Figure 47 is essentially a special view of a *fold*,[33,39] and arises in the buckling of engineering frames.[36,170,171]

The stable-symmetric point of bifurcation shown in Figure 48 is a *stable cusp* and has just been seen in the response of an elastic column. It also arises in the distinct buckling and post-buckling of elastic plates which are quite stiff and structurally useful in the post-buckling range in contrast to the rather flexible column.

The unstable-symmetric point of bifurcation shown in Figure 49 is an *unstable cusp*, and has also just been seen in the buckling of a deep arch. It arises extensively throughout the buckling of thin elastic shells, where very severe imperfection sensitivities can be encountered, in contrast to the rather mild ones observed in frames and arches. It is this mildness that allowed Roorda to make his

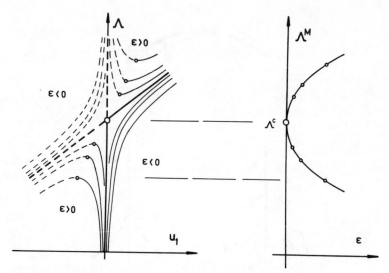

Figure 47 Canonical diagram of an asymmetric point of bifurcation showing load-deflection curves and the imperfection sensitivity

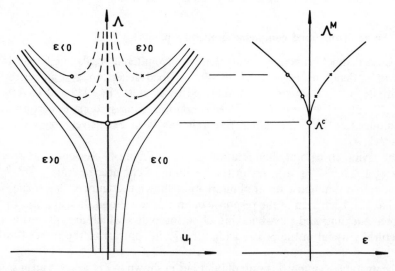

Figure 48 Canonical diagram of a stable-symmetric point of bifurcation showing load-deflection curves and the 'imperfection sensitivity' of the complementary minima

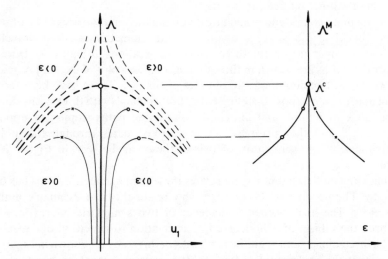

Figure 49 Canonical diagram of an unstable-symmetric point of bifurcation showing load-deflection curves and the imperfection sensitivity

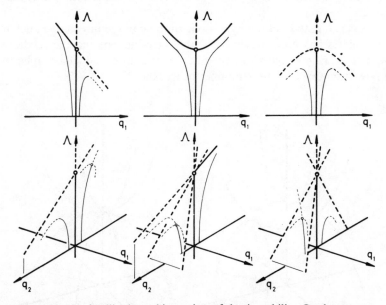

Figure 50 Six familiar branching points of elastic stability. On the top row are the three distinct points, the asymmetric, the stable-symmetric and the unstable-symmetric. On the bottom row are the three semi-symmetric twofold points, the monoclinal, the homeoclinal and the anticlinal

60

celebrated experimental studies of branching and imperfection sensitivity for these three distinct bifurcation points.[27]

Moving on to compound branching involving the simultaneity of two or more distinct buckling loads, three common semi-symmetric two-fold branching points are compared with the three distinct bifurcations in Figure 50. Here the equilibrium paths are shown in three dimensions on a plot of the load Λ against the two active buckling mode amplitudes, q_1 and q_2. These have been termed monoclinal (with one post-buckling path), homeoclinal (with three paths sloping in the same direction), and anticlinal (with three paths sloping in opposite directions),[39] and they arise by taking various routes through the umbilic catastrophes.[70] The generation of two of these can be seen in Figure 20 of Chapter 1.

A buckling model that neatly illustrates these semi-symmetric forms has been given by Thompson and Gaspar,[172] who relate them to Zeeman's umbilic bracelet.[173] The higher-order coalescence of two symmetric bifurcations has also been the subject of a model study[87] in relation to structural optimization, and the three-dimensional equilibrium paths of this study are shown in Figure 51. We see that the coalescence of two stable-symmetric branching points is here generating unstable coupled behaviour that will lead to an unexpected imperfection sensitivity. We should note here that theorems due to Duen Ho[174–176] have proved useful in identifying the most severe types of imperfection in compound buckling.

Hunt's elegant study of the hyperbolic umbilic in the interactive buckling of stiffened plates has been presented in the introduction, and the relationships between the elastic stability of conservative structures and catastrophe theory have now been explored by a number of workers.[34,39,177–182]

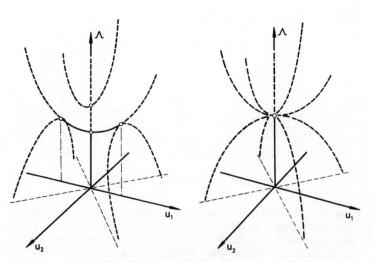

Figure 51   The coalescence of two stable-symmetric bifurcations showing the drawing-in of unstable coupled paths

Poston and Stewart, in their definitive treatise on catastrophe theory,[11] have looked in detail at the complete unfolding of a simultaneously buckling plate and have identified the appropriate *physical* unfolding parameters, while Schaeffer and Golubitsky[183] have looked at the role of boundary conditions on mode jumping in the post-buckling of rectangular plates. A valuable recent contribution by Chillingworth looks at universal bifurcation problems in solid mechanics.[184]

The continuum background of elastic stability theory can be seen in the definative contribution of Knops and Wilkes to the *Handbuch der Physik*.[185]

2.9 Practical problems of strength estimation

We shall not attempt here to survey in detail the extensive technical literature on the buckling and post-buckling of practical engineering structures, but rather give some significant and modern references to guide the interested reader.

Starting with the classic studies of Kármán and Tsien[186] and Koiter,[17] the erratic and explosive buckling of thin elastic shells has been a major preoccupation of designers in the aero-space field. Such shells have a very severe imperfection sensitivity, often associated with simultaneous, or near-simultaneous buckling behaviour. An example is shown in Figure 52 for the post-buckling imperfection sensitivity of a cylindrical shell as analysed by Koiter. This

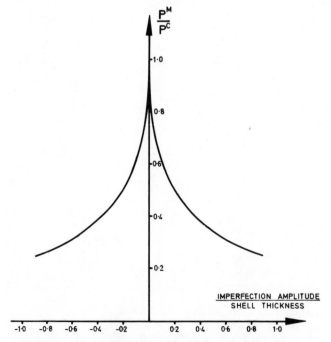

Figure 52 The imperfection sensitivity of an axially compressed cylindrical shell due to Koiter

shows the variation of the failure load as a fraction of the critical load of a perfect shell, against the amplitude of an assumed geometrical imperfection in the shape of the middle surface for a complete cylindrical shell subjected to uniform axial compression.

Following a seminal paper by Van der Neut,[187] much recent activity has focused on the interactive buckling of stiffened plates and shells,[188-195] while the extensive work of Croll and co-workers has aimed to develop design approaches for a wide variety of stiffened and unstiffened thin-walled structures.[196-199] Numerical methods of analysis based on general stability studies have been given by a number of workers,[200-203] and a design-oriented analysis of trusses should be noted.[204] Creep buckling has been surveyed by Hoff,[205] while the influence of creep on the post-buckling response of structures has recently been studied by Hayman.[206] Excellent reviews of buckling and post-buckling theory have been made by Koiter, Budiansky, and Hutchinson,[207-210] while useful modern engineering references on theory and practice are due to Gioncu and Ivan,[211,212] Rhodes and Walker,[213] and Allen and Bulson[214].

CHAPTER 3

Astrophysics and Gravitational Collapse

Many of the ideas and concepts of stability theory arose historically in astrophysical problems of stars and planets. The modelling of these as solid or fluid continua generates boundary-value problems very similar in form to those encountered in engineering mechanics, and the high degrees of symmetry involved allow considerable scope for analytical solutions.

We consider first the gravitational collapse of a massive cold star, which is a unique *mechanics* problem in Einstein's *general* theory of relativity: the latter theory must be employed, not because of high velocities[215] but because of the *large-scale curvature of space-time*.[216]

We next consider the *thermodynamical* instabilities of a hot spherical stellar system, which suffers what Lynden-Bell has called the *gravothermal catastrophe*.

We finally take a brief look at the progressive instabilities of a *rotating* fluid mass which is held together by its self-gravitation. This problem has deep historical ties with the genesis of bifurcation theory[4, 217] and is relevant to the formation and evolution of the planets.

3.1 Collapse of a massive cold star in general relativity

To shed some light on the realistic gravitational collapse of a massive hot star, which would include the complications of angular momentum, magnetic fields, turbulence, and shock waves, Harrison, Thorne, Wakano, and Wheeler[218] consider the ground state of a system of A baryons (neutrons and protons) that have been catalysed to the end point of thermonuclear evolution and *cooled* as closely as desired to absolute zero.

For $A = 1$ the ground state corresponds to one hydrogen atom, for $A = 4$ to a helium atom, and for $A = 56$ an iron atom. Stepping up to $A = 56 \times 10^6$ the catalysed end product is a set of 10^6 iron atoms, Fe^{56}, arranged in a body-centred cubic lattice. When the baryon number reaches the order of 10^{56} or 10^{57} the self-induced gravitational forces are so powerful that the extreme pressure raises the electrons to relativistic energies and they transmute protons to neutrons. The nuclear composition changes from Fe^{56} to heavier and more neutron-rich nuclei.

Using the unique and universal equation of state corresponding to this cold catalysed matter, an analysis is made of the spherically symmetric equilibrium configurations of a self-gravitating stellar mass, the necessary general relativistic

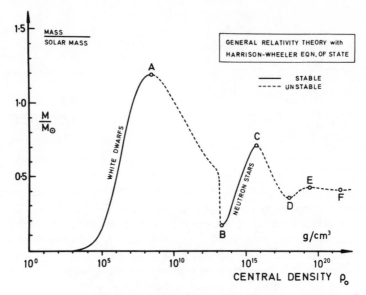

Figure 53   Equilibrium solutions for a cold stellar mass on a plot of the mass energy against the central density

equation of hydrostatic equilibrium being obtained by extremizing the mass as sensed externally. The stability of these equilibrium configurations against radial spherically symmetric disturbances is determined by studying the characteristic acoustical modes of vibration and by examining the second variation of the mass energy for a fixed baryon number.

The results of these equilibrium and stability analyses are summarized in Figure 53 in which the mass energy is plotted against the central density, and we see that there are only two regions of stability corresponding to white dwarf stars and neutron stars respectively. With an increasing number of baryons and mass energy each of these regions terminates at a critical equilibrium state, or fold catastrophe, at which gravitational forces dominate and collapse can start. Below these critical points collapse is only prevented by a potential barrier that tends to zero as the critical state is approached, and quantum mechanical leakage through this diminishing barrier is of course increasingly possible.

We note in particular that the critical masses are of the order of the mass of the sun, and that the critical equilibrium states beyond C all represent a further *destabilization* of the stellar system.

In an important secondary analysis that departs from the equation of state of cold catalysed matter, it is shown[218] that 'no equation of state which is compatible with causality and with freedom of matter from microscopic collapse can save a system from having a configuration which is unstable against collective gravitational collapse'. Harrison *et al.* finally discuss the *idealized* dynamics of the collapse during which the total mass energy, including the relativistic equivalents of both potential and kinetic energy, remains constant. Since no suitable stable equilibrium states seem to be available, it is felt that the loss of stability at the critical equilibrium states A and C represents the beginning of collapse towards a black hole in space, a collapse that could be triggered by the steady gravitational capture of matter by a cold white dwarf or neutron star.

Historically the existence of the first two peaks terminating the stable domains can be inferred from Landau's pioneering order of magnitude discussion of 1932.[219] The first peak, at which the electron pressure is overwhelmed, was analysed in a limited way by Chandrasekhar[220,221] in 1935 using a Newtonian equation of hydrostatic equilibrium to obtain an asymptotic approach to the critical mass. The second peak was analysed by Oppenheimer and Volkoff[222] in 1939 using a general relativity equation of equilibrium for an ideal neutron gas.

The final complete curve, showing for the first time both of the aforementioned peaks on a single path, was determined by Harrison et al.[218] These workers used the Harrison – Wheeler equation of state of cold catalysed matter in which there is a uniquely determined nuclear species dominant at any given pressure. This equation of state was employed with the Tolman – Oppenheimer – Volkoff general relativity equation of hydrostatic equilibrium which was integrated numerically starting with a chosen value of the central density. In a separate analysis these workers also showed that there is an infinite number of critical points (folds) as the mass becomes a damped periodic function of Log ρ_0 for high values of the central density ρ_0.

As the equilibrium path emerges from the origin we start with an atom of hydrogen, a cannon ball of iron, objects of planetary mass, and finally the cold white dwarfs. The first instability at the fold A, which can be predicted using only a Newtonian equilibrium equation, corresponds to the overwhelming of the electron pressure at a baryon number of 1.4×10^{57}. At the minimum B, the stellar matter has been crushed to a substantial fraction of nuclear density and its increased rigidity leads to the stable neutron stars. At the second peak C the gravitational forces finally dominate even this nuclear rigidity. We note here that the baryon number at C, 0.84×10^{57}, is lower than that at A. So if in an idealized dynamical collapse of a cold white dwarf there was no change in the baryon number (no matter thrown off) the star could not stabilize as a neutron star.

We can note here that a figure of 10^{57} atoms per star is a fairly typical number (assuming 10^{11} stars in a typical galaxy and estimating 10^{10} galaxies in the universe, this gives the atoms in the universe as 10^{78}). It is therefore not unusual for a star to collapse, at least to a neutron star, when it cools after its nuclear fuel is exhausted.

The stability of this equilibrium sequence was first investigated by Misner and Zapolsky[223] in 1964 using a vibration analysis based on a variational principle of Chandrasekhar. They showed that the stability in the first normal mode of vibration is lost as the path passes through the second peak C, but that the path in the region of E is everywhere unstable with respect to this first mode. They thus established the stability of region BC and showed that the stability in mode one is *not* recovered at the minimum D, which therefore must represent a further loss of stability.

In fact, as we shall see, each subsequent maximum and minimum represents a further destabilization, which is explained by Harrison *et al.* as follows:

> For any given high central density there is an inner region of the star which is teetering on the verge of collapse. An acoustical mode of *high* order has many alternating zones of density increase and density decrease inside this critical region. Therefore, it has no effective coupling to the decisive degree of freedom. To excite this mode is to excite a vibration, not to promote a collapse. On the other hand, to excite an acoustical mode of *low* order is to make a density increase throughout the critical region, and to initiate collapse: hence the instability of the acoustical modes of lower order. The demarcation between unstable modes and stable modes therefore comes at that stage at which the location of the innermost nodal sphere in the vibration first penetrates well into the critical sphere whose radius varies as $\rho_0^{-1/2}$.

The complete stability analysis for spherically symmetric disturbances was given by Harrison *et al.*[218] They identified the mass energy M as a governing *potential* determining both the equilibrium and stability with the baryon number A (or equivalently the mass before assembly M_A which is proportional to A) playing the role of a *control parameter* and the central density ρ_0 playing the role of an active *generalized coordinate*. They draw a three-dimensional picture of the energy transitions in the $M - A - \rho_0$ space exactly equivalent to that of the fold catastrophe with $\partial M / \partial A = V' \neq 0$ at the critical point. Under this condition they observe that both the potential M and the control A take extreme values on the equilibrium path at the critical point, and they demonstrate the two-thirds power-law cusp in the energy-control projection.[47] This is illustrated later in Figure 56.

Now the mass before assembly M_A is simply the baryon number A multiplied by the standard baryon mass μ_S and so represents a valid control parameter. It is, however, closely equal to the equilibrium mass energy M, and for this reason the projection of the equilibrium path onto the energy-control plane lies very close to the 45° line. To observe the cusps it is therefore more convenient to plot $M - M_A$ (which is of course also a valid potential function) against the control M_A, and we have done this in Figure 54. On this picture, by inspecting the energy levels close to the folds, we can deduce the sequence of stability transformations; these are summarized in Figure 55.

At A the originally stable path loses its stability with respect to mode one by the passage through zero of the first stability coefficient V_{11}. Stability is recovered at B by the passage back through zero of V_{11} but is finally lost at C as this stability coefficient again becomes negative. Each fold after C now corresponds to a loss of stability, V_{22} becoming negative at D, V_{33} becoming negative at E, V_{44} becoming negative at F, etc.

Looking closely at Figure 54 we see to our surprise that some of the equilibrium configurations have positive values of $M - M_A$ so that their mass

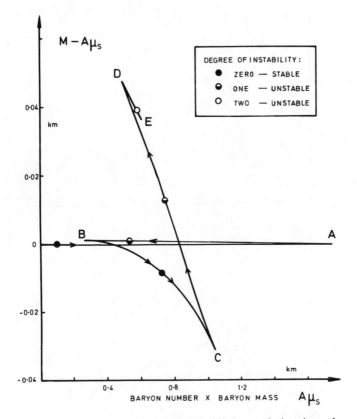

Figure 54 The projection of the equilibrium solution into the energy-control space, showing the cusps at the folds

energy is greater than the mass before assembly: 'There thus exist solutions of the general relativity equation of hydrostatic equilibrium which have excess energy relative to dispersed Fe^{56} atoms at rest at infinite separation.'[218] The states are thus *metastable* against complete dispersion. Harrison *et al.*, following Zel'dovich, describe how these states of excess energy could be achieved by a hypothetical pumping operation involving the addition and removal of baryons to induce continuous motion along the equilibrium path.

We notice also that $M - M_A$ takes its absolute lowest value at the critical point C so that we have Theorem 14 of the above:[218] 'The binding at the critical point C is the tightest binding which can be achieved, according to the Harrison – Wheeler equation of state, in any equilibrium configuration of cold catalysed matter held together by its own mutual gravitation.'

These statical computations and the associated linear vibration analyses form an attractive illustration of a sequence of fold catastrophes, and the slow gravitational capture of mass by a white dwarf or neutron star gives us the slow evolutionary change of the control parameter. However, once collapse is initiated

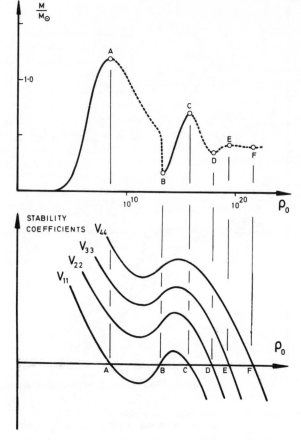

Figure 55 The variation of the stability coefficients shown schematically against the central density

at A or C towards states of greater density most of the assumptions of the highly idealized analysis break down. Both baryons and radiant energy may be lost from the system and so the value of the control parameter is not preserved. Questions such as the possible formation of a black hole by the collapsing star are thus beyond the scope of this idealized study.

We must finally comment on the close numerical equivalence of M_A and M on the equilibrium path, since this has caused some misunderstanding in the past. From a conceptual point of view the distinction is clear and vital, as has been emphasized by E. C. Zeeman in private correspondence with the author: the mass *before assembly*, M_A, is the catastrophe theory *control* parameter while the mass *after assembly*, M, is the catastrophe theory *potential*. However, when we have solved the equilibrium equations and come to plot the equilibrium values of M and M_A against the generalized coordinate ρ_0 we shall find that the two graphs are for all practical purposes identical. To illustrate this point we show in Table 1

Table 1 A table showing the extreme ratios of the mass energy M to the mass before assembly M_A, corresponding to the critical equilibrium states

Control parameter — mass *before* assembly M_A Potential function — mass-energy *after* assembly M		$\dfrac{M}{M_A}$
First maximum	A	0.9999
First minimum	B	1.0041
Second maximum	C	0.9701
Second minimum	D	1.0980
Third maximum	E	1.0603

M_A = baryon number A × baryon mass μ_S

the ratio of M to M_A at the critical points where it departs furthest from unity, and we see that the equilibrium graphs could not easily be distinguished.[47]

Further details of the tantalizing theoretical problem of gravitational collapse can be found in the authoritative treatises of Wheeler,[224,225] Weinberg,[226] Hawking and Ellis,[227] Misner, Thorn, and Wheeler,[228] and Rees, Ruffini, and Wheeler.[229]

3.2 Supernovae explosions of cooling stars

The collapse of a growing neutron star towards a black hole is still a somewhat speculative event. In contrast the collapse of a massive cooling white dwarf to a neutron star is comparatively well understood, giving rise as it does to a spectacular *supernova*.

Thus when a *massive* isolated star begins to cool at the end of its steady thermonuclear evolution, the internal temperature and pressure fall and the core begins to collapse under the force of its own gravity. The resulting implosion sets up huge disturbances in the outer layers of the star, which are then ejected at thousands of kilometres per second. Meanwhile the core collapses violently from thousands of kilometres to tens of kilometres in a fraction of a second, to create a highly compressed and exceedingly dense *neutron* star. On the surface of such a star an apple would weigh approximately 30 million tonnes, and an object dropped from a table would hit the floor at a speed of 2 million miles per hour.

Any small rotation of the original star will be magnified in this central neutron star due to the enormous decrease in its moment of inertia, and its fast rotation will make it identifiable as a *pulsar* emitting regular pulsed radio signals.

During this supernova event, the luminosity will for a brief period equal that of a million normal stars, and over the following decades the expanding envelope transforms itself into a glowing supernova remnant, loosely described as a nebula, the latter being originally used to describe all hazy astronomical objects including galaxies, star clusters, and gaseous clouds.

The classic example of such a supernova is the Crab Nebula in Taurus, which with its central spinning pulsar is a key object in modern astrophysics. It is the first object to be listed in Messier's catalogue, and is therefore designated as M1.

The Crab Nebula is now known to be the remnant of a supernova collapse that was observed as a 'guest star' by Chinese astronomers in 1054. It was then bright enough to be seen in the daytime for 23 days, and was visible at night for a total of 653 days, being now invisible to the naked eye. It is easily visible today with a small amateur telescope or binoculars, and in a large instrument it is seen as a spectacular tangle of luminous filaments as in Photograph 3a. It was clearly a supernova, rather than the distinctive nova event, and being the most intense radio source in the sky, apart from the sun, it is of considerable interest to amateur radio astronomers today.

The central spinning neutron star, namely a pulsar, was discovered in the Crab Nebula on 9 November 1968. The period of 33 milliseconds implies that the neutron star is spinning at 30 revolutions per second, a phenomenal rate for such a massive object. The pulsing of the Crab Nebula neutron star is remarkably regular, and can be observed over the entire electromagnetic spectrum from 100 MHz, through the optical range, as far as the X-rays. As with other pulsars, a slight slowing down has been detected. The total mass of the Crab Nebula has been estimated at two to three solar masses, of which the central neutron star accounts for about one solar mass.

Over a number of years, the details of the Nebula can be observed to be moving radially outwards from the centre, and projecting backwards in time gives good correlation with the Chinese date. The expansion of the Nebula is, for example, cleverly indicated in a composite picture due to Virginia Trimble who combined a positive photographic print of 1950 with a negative print of 1964. On such a montage, fast-moving filaments have a black leading edge.

About 10 supernovae are observed in galaxies other than our own Milky Way each year, and over 400 extra-galactic supernovae have now been catalogued. Supernovae of the crab type are in fact crucially important for the production of heavy elements in the galaxies.

Curiously, no supernovae have been observed in the Milky Way since the invention of the telescope, although statistically one is long overdue. Some significant supernovae in our galaxy are that of 1006 described by Oriental and Arabic sources, the Crab of 1054, Tycho Brahe's sighting of 1572, and Kepler's supernova of 1604. A collapsed star core within Tycho's supernova has in fact been recently claimed in *New Scientist* of 25 September 1980. New high-resolution observations with the Cambridge Five Kilometre radio telescope by Steve Gull and Guy Pooley reveal the details of Photograph 3b. This picture is formed from their radio observations, and shows the remnants of Tycho's supernova with a minute but intense radio source near to, though not quite at, the

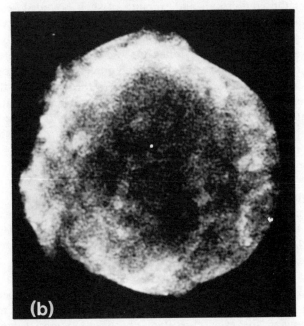

Photograph 3. Remnants of supernova instabilities. (a) The Crab Nebula, the brilliant explosion of which was recorded by ancient Chinese astronomers in 1054. A central spinning neutron star, a pulsar, was discovered in 1968. Photograph by courtesy of the Hale Observatories. (b) The remains of a supernova observed by Tycho Brahe. This new radio photograph taken in 1980 at wavelength 11 cm shows a collapsed neutron star near the centre as reported in *New Scientist* of 25 September 1980. Reproduced with the permission of S. Gull and G. Pooley

centre. This is *suggested* to be a neutron star produced in the supernova, although no pulsations have yet been observed—in contrast to those detected from the neutron star in the Crab Nebula.

3.3 Gravothermal catastrophe of hot stellar systems

The thermodynamic stability of self-gravitating stellar systems provides a nice illustration of fold catastrophes. Here there is a well-defined thermodynamic potential, such as the negative of the entropy for an isolated system, and other quantities such as the energy E or the radius R can be viewed as control parameters.

Attention is here restricted to isothermal spheres of identical particles, and the onset of the so-called gravothermal catastrophe is predicted using a conjugate theorem. This thermodynamic gravitational instability associated with the onset of a core-halo or red-giant structure is shown by Lynden-Bell and Wood[230] to throw light on the evolution and final fate of *galaxies*, stellar *clusters*, and possibly individual *stars* themselves.

Lynden-Bell and Wood give in their paper a thorough thermodynamical discussion of the equilibrium and stability of such bounded self-gravitating isothermal spheres. They observe that the thermodynamics employed is unusual because self-gravitating systems have *negative specific heats*, so if heat is allowed to flow between two of them the hotter one loses heat and gets hotter while the colder one gains heat and gets even colder. Evolution is therefore often away from equilibrium as we shall see. Their work serves to throw light on the results of numerical experiments by Aarseth[231] and the problem of Antonov[232] by showing that the thermodynamics of concentrated systems makes them evolve into core-halo structures. They discuss in particular the loss of stability of an equilibrium path at a limit point.

The stellar system considered has fixed mass M and is imagined to be held in a container of volume V exerting a pressure P. If the container is *rigid*, V is a prescribed quantity while P is the passive constraining pressure. If, on the other hand, the container is infinitely flexible and surrounded by a prescribed *dead* pressure P, this pressure becomes the controlled quantity, under which the system can choose its own volume V. We have here a direct analogy with the dead and rigid loading of engineering structures.[47]

The wall of the container can, moreover, be either a perfect thermal insulator or a perfect conductor linking the system to the constant temperature of a surrounding heat bath. The resulting four-fold classification is summarized in Table 2, showing for each case the appropriate potential function. This classification is the same as that of Glansdorff and Prigogine (Ref. 233, page 46) for non-gravitating systems.

Focusing attention on the isolated rigid case, Lynden-Bell and Wood take Boltzmann's expression for entropy and maximize it subject to the constraints of constant M and E using Lagrange multipliers to obtain the differential equation of equilibrium. They repeat Antonov's proof that only spherically symmetric

Table 2 Thermodynamic potentials for the equilibrium and stability of a system of fixed mass M, from Glansdorff and Prigogine[233] and Lynden-Bell and Wood[230]

System	Controls	Potential function
Thermally isolated		
Rigid constraint	Energy E Volume V	Negative of the entropy, $-S$
Dead pressure	Entropy S Pressure P	Enthalpy H
Heat bath		
Rigid constraint	Temperature T Volume V	Helmholtz free energy F
Dead pressure	Temperature T Pressure P	Gibb's free energy G

states can correspond to local entropy maxima, so that only spherically symmetric solutions need be considered, and specialize the differential equation to the well-known form for the isothermal gas sphere. They finally establish the extremal energy criteria of our table. Using these formulations they then determine the loss of stability of isothermal spheres for the four distinct cases of Table 2. In each case the loss of stability is associated with a fold or limit point on the equilibrium path.

The fold for the isolated rigid case is drawn in full with a sketch of the entropy transformation corresponding to our drawing[47] of Figure 56. We should notice,

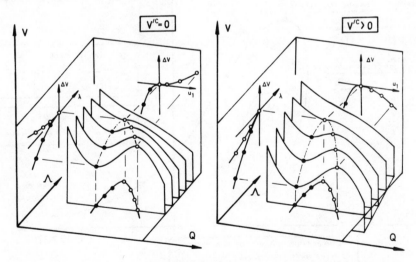

Figure 56 Three-dimensional views of a fold catastrophe showing the potential energy V as a function of the generalized coordinate Q at different values of the control parameter Λ. The right-hand picture shows the tilting of the energy-control cusp when $\partial V/\partial \Lambda|^c = V'^c \neq 0$

however, that in their drawing (Ref. 230, Figure 2, page 509), since they draw contours of constant entropy (rather than plots of S at different values of the control), *maximum* values of S at a fixed control value correspond to a contour exhibiting a *minimum* at that point.

Lynden-Bell and Wood also sketch equilibrium surfaces and discuss briefly the possibility of a loss of stability at a perhaps unidentified branching point.

Katz[234-236] draws on these calculations to show how a conjugate theorem can be used to establish the stability transitions at the *succession* of folds along the equilibrium path. We have established this theorem in detail for the structurally stable fold,[47] but it is worth noting that Katz proves it with some relaxation of our continuity requirements.

To make his stability predictions, Katz considers the plot of the inverse of the temperature, T^{-1}, against the energy E, shown in Figure 57, and focuses attention on the *rigidly constrained* system. On this graph the equilibrium curve is parametrized in terms of the *density contrast*, that is the ratio, h, between the density in the centre and the density on the boundary. This parameter is seen to increase as the curve spirals inwards.

Now for high temperatures for which T^{-1} tends to zero, the stellar system behaves like a stable collisionless gas, so the region of the curve to the right is known to be stable. Thus we have a region of known stability from which progressive losses of stability can be traced.

For a rigid stellar system that is *thermally isolated* the changes of stability occur

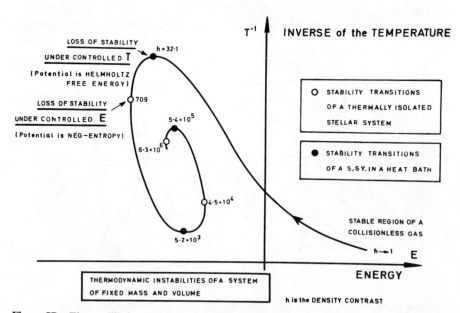

Figure 57 The equilibrium path of a stellar system, parametrized by the density contrast h. The curve is established for isothermal spheres of identical particles: the mass and volume are fixed, and the stability transitions are as indicated

at the vertical tangents where the controlled energy E is stationary: and Katz uses his conjugate theorem to show that as the curve spirals inwards, *degrees of stability* are progressively *lost*, with no restabilization. Similarly, for a rigid stellar system that is in a *heat bath* the changes of stability occur at the horizontal tangents where the controlled temperature T is stationary: and Katz again uses the conjugate theorem to show that as the curve spirals inwards there is *again* no restabilization.

We see that for low T and low E there are no available equilibrium states, and it follows that for a single isothermal sphere within a non-conducting box, no equilibrium states of energy E (< 0) and mass M exist once the radius of the box is greater than $0.335GM^2/(-E)$, where G is the gravitational constant.

3.4 Core-haloes in globular clusters

These thermodynamic studies suggest an evolution towards the core-halo structure displayed by *globular* star clusters, a nice example of which is shown in Photograph 4. The thermodynamic studies are backed up by star cluster simulations on a computer, which reveal some of the core-halo features. In this connection we would mention the pioneering work of Aarseth using strict N-body techniques. Such computer analyses can get nearer to real star cluster conditions by including the complications of mass loss, different stellar masses (rather than identical particles), and external perturbations such as a tidal field.

An example of one such simulation, made by Terlevich in association with Aarseth, is shown in Photograph 5. Here a thousand stars of different masses were started with random velocities and distributed inside a sphere, the area of the plotted circles being proportional to the mass of the star represented. Views (a) and (b) show the cluster at age 463 million years from two different directions to illustrate the compression due to the galactic tidal field. View (c) shows the same cluster after 703 million years when only about 200 stars remain. In addition to the simple internal gravitational interactions, this N-body simulation includes the effects of the galactic tidal field, mass loss from simulated supernova events in the star populations, and interstellar clouds. A spectacular movie made from this computer model shows the stars shuttling about in the cluster like molecules in a hot gas as millions of years tick away.

View (d) shows for comparison a real *open* star cluster, a larger number of stars being needed to simulate a *globular* cluster. One important feature of such simulations can be the appearance of central binaries that absorb a large fraction of the total energy.

3.5 Evolution of rotating planetary masses

The relative equilibrium and stability of a rotating mass of a homogeneous self-gravitating liquid is a classical problem in astrophysics, with its roots dating back to Poincaré[4-6] and even Newton. It is discussed extensively in the treatises of Darwin,[237] Jeans,[238,239] Lyttleton,[240] Ledoux,[241] and Chandrasekhar,[242] and

Photograph 4 A globular star cluster in Hercules, Messier number M13, showing the core-halo structure. The gravothermal collapse of globular cluster cores is discussed by Lynden-Bell and Eggleton (*Mon. Not. R. Astr. Soc.*, **191**, 483, 1980) who touch on the possible formation of central black holes in stellar systems. Photograph by courtesy of the Hale Observatories

we consider here the system under the condition of free rotation that might pertain to a primitive planetary mass.

Since the essential question is one of *relative* equilibrium and stability, it is normal to study rotating systems in a frame of reference rotating with the body at an angular velocity W. In conditions of free rotation we could, for example, choose the reference frame such that the angular momentum relative to this frame remains zero; W would then vary with changes in the moment of inertia I.

To make our discussion more concrete, let us first focus attention on *ellipsoidal* states of relative equilibrium of a rotating liquid mass, following Katz.[235] We assume that we have uniform density ρ and semi-axes a, b, and c that are rotating

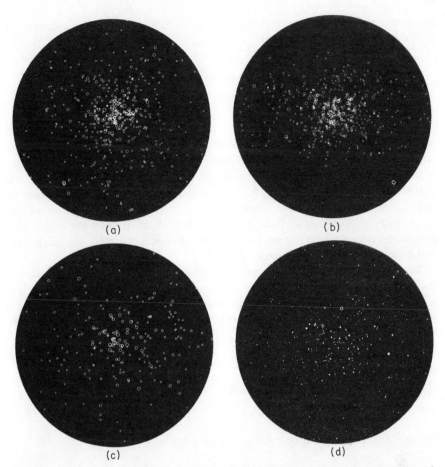

(a) (b)

(c) (d)

Photograph 5 Three stills from a computer simulation of a star cluster, plus one real cluster: (a) and (b) show the cluster at age 463 million years from two different directions to illustrate the galactic tidal compression while (c) shows the same cluster after 703 million years when more stars have escaped. These pictures were obtained from a dynamic computer model of a *galactic* star cluster in which the individual star orbits are calculated by the direct N-body method. The calculations were performed by Mrs E. Terlevich at the Cambridge Institute of Astronomy and were kindly supplied by Sverre Aarseth. Similar N-body simulations of Aarseth (*Astrophys. and Space Sci.*, **13**, 324, 1971) show the spectacular evolution of a highly energetic central binary. The actual star photograph (d) shows the *open* or *galactic* cluster M67. Reproduced with the permission of Mrs E. Terlevich

around axis c with angular velocity W. The mass

$$M = \tfrac{4}{3}\pi abc\rho$$

and the angular momentum

$$L = \tfrac{1}{5}M(a^2 + b^2)W = IW$$

are conserved, and in co-moving coordinates along a, b, and c the effective potential is given by Ledoux as

$$V = \tfrac{1}{2} \int \rho \varphi \, dv + \frac{L^2}{2I}$$

where φ is the gravitational potential.

The conditions of elementary catastrophe theory are now fullfilled because equilibrium configurations are given by stationary values of V and *secular* stability holds if V is a local *minimum*. Here, by secular stability, we mean stability in the presence of complete internal dissipation, the effect of which is to cancel any gyroscopic stabilization arising from coriolis forces. We can quote a theorem from Ziegler:[20] 'Dissipative forces, applied to other than non-gyroscopic conservative systems, may have a destabilizing effect. If they are added to a gyroscopic conservative system, and if the dissipation is complete, they cancel the stabilizing effect of the gyroscopic forces.'

Restricting attention to the ellipsoids, we can regard a, b, c, and W as variables with the two constraints, M and L constant, and we could regard the density ρ as a single control parameter. In physical terms we might, for example, be thinking of an isolated primitive planetary mass whose density is *slowly* increasing due to cooling.

The constraints can be eliminated by reducing the number of coordinates to two, and we write them as

$$q_1 = \left(\frac{a}{c}\right)^2 \qquad \text{and} \qquad q_2 = \left(\frac{b}{c}\right)^2.$$

We can also introduce a *composite* control parameter

$$\Lambda = 25\left(\frac{4\pi\rho}{3M}\right)^{1/3} \frac{L^2}{3GM^3}$$

where G is the gravitational constant, and our energy function has the final form

$$V = \frac{3}{10} GM^2 \left(\frac{4\pi\rho}{3M}\right)^{1/3} \left[-(q_1 q_2)^{1/6} \int_0^\infty \{(1+\lambda)(q_1+\lambda)(q_2+\lambda)\}^{-1/2} d\lambda \right.$$

$$\left. + \Lambda \frac{(q_1 q_2)^{1/3}}{q_1 + q_2} \right].$$

The equilibrium equations

$$\frac{\partial V}{\partial q_1} = \frac{\partial V}{\partial q_2} = 0$$

have two sets of solutions, the Maclaurin spheroids with $a/c = b/c$ and the Jacobi ellipsoids with $a \neq b$. The latter bifurcate from the former at the point $a/c =$

$b/c = 1.716$, $\Lambda = 0.769$. For these ellipsoidal solutions, Katz uses Thompson's technique[47] to prove the stability of the Jacobi ellipsoids everywhere and the Maclaurin spheroids for Λ less than 0.769.

More complete calculations, allowing a departure from the ellipsoidal states, show that the Jacobi ellipsoids later become unstable against a pear-shaped form, and the corresponding secondary branching points are shown in Figure 58. Here an angular momentum parameter is plotted against two deformation variables $(a - b)/c$ and p. Here p denotes displacement in the third, non-ellipsoidal, pear-shaped mode, being zero for the Maclaurin and Jacobi series. The graph has been plotted from Lyttleton's numerical results for deformations in the ellipsoidal

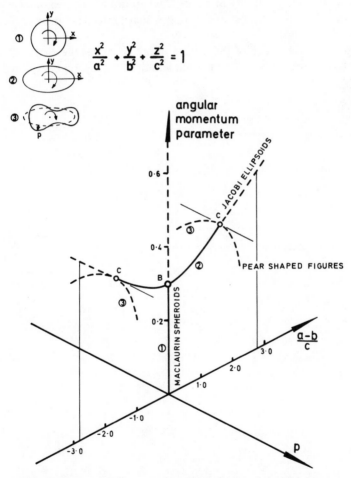

Figure 58 The instability of a primitive planet. This picture shows the progressive loss of stability and symmetry of a self-gravitating rotating liquid mass. Similar instabilities may play key roles in the development of planetary systems

modes, but variations with respect to p are merely schematic, the calculations at this stage becoming so complicated that they admit only qualitative predictions.

When the angular momentum parameter is small there is just one equilibrium path lying along the control axis with $a = b$. This is the Maclaurin series of oblate spheroids, and we see that the series continues to infinite momentum where we have an infinitesimally thin, infinitely wide disk. This equilibrium path does, however, become unstable at the stable-symmetric point of bifurcation B, an example of a stable cusp, where it intersects the Jacobi series of ellipsoids for which $a \neq b$.

This series also continues to infinite momentum, where the liquid mass now has the form of an infinitely long line with $a = \infty$ and $b = c = 0$. However, this Jacobi series loses its initial stability at states C where we have a secondary unstable-symmetric bifurcation or unstable cusp, at which it intersects an unstable series of the pear-shaped form.

The possible relevance of these instabilities to planetary evolution and the origin of satellites is discussed at some length in the references quoted.

CHAPTER 4

Bifurcational Instability of an Atomic Lattice

If a perfect crystalline solid is pulled in a tensile test along an axis of lattice symmetry, the crystal might be expected to elongate in the direction of the applied force, leading eventually to a direct tensile fracture. Under certain conditions, however, an unexpected symmetry-breaking shearing deformation can arise, giving a premature shearing failure akin to the formation of a shear band.[243]

We shall present in this chapter a recent investigation of this bifurcational phenomenon due to Thompson and Shorrock[82] with the kind permission of Pergamon Press. This will allow us to introduce many fundamental ideas of bifurcation theory in an interesting and specific setting.

4.1 Symmetry-breaking phenomena

Consider the sheet of close-packed atoms shown in Figure 59 pulled in uniaxial tension by the direct tensile stress σ_{11}, and suppose for the time being that there is no shearing stress σ_{12}. We would then expect the crystal to elongate in the direction of σ_{11} with a direct strain ε_{11}, the stress–strain curve eventually reaching a maximum as the atomic bonds are overcome. The corresponding *primary* equilibrium path is shown in the plane of σ_{11} and ε_{11} in the top diagram.

This primary symmetric path of a perfect crystal can, however, become unstable at a branching point A as shown by the linear eigenvalue analysis of Macmillan and Kelly,[244] and we have demonstrated on the basis of a non-linear analysis[82] that this is associated with an *unstable-symmetric* point of bifurcation or *cusp catastrophe*. A secondary path of unstable sheared states with non-zero shearing strain ε_{12} thus emerges from this point as drawn, and the bifurcation load would represent the failure stress of our perfect crystal. At the bifurcation load, even an infinitesimal disturbance would trigger a violent fracture of our mechanically stressed crystal.

A small shearing stress σ_{12} would clearly act as an initial imperfection, destroying the simple primary solution and rounding off the bifurcation as shown by the lighter lines. The projection into the $(\sigma_{11}, \varepsilon_{12})$ plane is shown in the lower left, and we see that we have an unstable cusp with the two-thirds power-law imperfection sensitivity of the lower right diagram. Here the cusp in the two-dimensional $(\sigma_{11}, \sigma_{12})$ control space represents a failure-stress locus. A similar

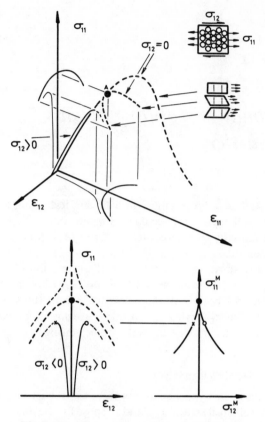

Figure 59 The upper picture shows the equilibrium paths of the atomic lattice when $\sigma_{22} = 0$, while the lower left-hand picture shows the projection of these paths into the $(\sigma_{11}, \varepsilon_{12})$ plane where we see the well-known unstable-symmetric point of bifurcation or cusp catastrophe. In general, stable regions of paths are drawn as solid lines while unstable regions of paths are drawn as broken lines. The lower right-hand picture shows the corresponding stability or catastrophe boundary as a failure-stress locus with the predicted two-thirds power-law sensitivity

sharp sensitivity to *geometrical* symmetry-destroying imperfections is of course also to be expected.

If we now consider the addition of a lateral direct tensile stress σ_{22}, as shown in Figure 24 of Chapter 1, we have a problem with three controlled stresses. Positive σ_{22} tends to inhibit the bifurcational instability and push the branching point A away from the origin towards the tensile maximum. Thus a critical value, σ_{22}^C, can be calculated for which the branching point A is coincident with the primary

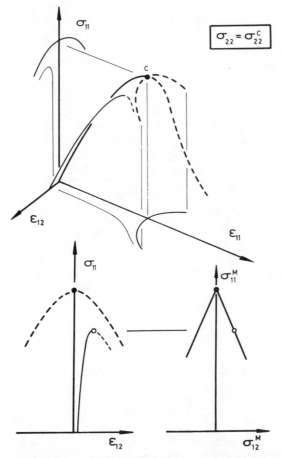

Figure 60 This shows how Figure 59 is modified by the addition of the direct stress $\sigma_{22} = \sigma_{22}^{C}$ which has shifted the point of bifurcation to the maximum of the primary stress–strain curve. The failure locus is now locally bilinear

maximum, as illustrated in Figure 60. Under this condition of $\sigma_{22} = \sigma_{22}^{C}$, we showed[82] that the original two-thirds power law in the two-dimensional control space of σ_{11} and σ_{12} is replaced by a bilinear failure locus as shown. The compound instability at the hill-top branching point is indeed seen, by Figure 20 of Chapter 1, to be an example of a hyperbolic umbilic catastrophe,[83] allowing us to sketch the complete three-dimensional failure locus in the space of σ_{11}, σ_{12}, and σ_{22} as shown in Figure 61. This can be compared with the cardboard model of the complete hyperbolic umbilic stability boundary of Photo 6 and the hyperbolic umbilic caustic of Photo 7.

Here we see that on the crest defined by $\sigma_{12} = 0$ the sharp cusp degenerates at the singularity corresponding to $\sigma_{22} = \sigma_{22}^{C}$ into two straight lines and thereafter

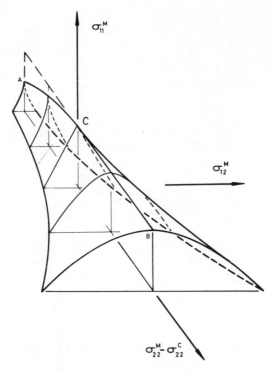

Figure 61 The stability boundary in the space of the failure stresses corresponding to the catastrophe boundary of the hyperbolic umbilic catastrophe

the curve has a parabolic form. This crest is curved between A and C where it corresponds to failure at a branching point, but it is approximately straight and parallel to the σ_{22} axis between C and B where failure is associated simply with the maximum of the primary equilibrium path: in this latter range the bifurcation has moved beyond this maximum and therefore no longer initiates the failure of the crystal. The failure surface has the characteristic form of the stability boundary of the hyperbolic umbilic catastrophe, which has often been likened to a breaking wave.

This attractive and concrete example of the hyperbolic umbilic catastrophe can be added to the imperfection sensitivity curve determined by Hunt (as summarized in Chapter 1) for the buckling of a stiffened panel as a second illustration of this catastrophe in non-linear elasticity.

Macmillan and Kelly have shown that the bifurcation at A can be up to 20 per cent. lower than that at the primary maximum. Often, though, it is much closer, and we might finally discuss the case in which the bifurcation at A is close to, but not strictly coincident with, the primary maximum. Here, the distinct cusp theory,[36] although valid very close to A, may give misleading results. In this circumstance it is safest to *fudge* or *compact* the analysis by *assuming* that the

Photograph 6 Photograph of a cardboard model of the stability boundary of the hyperbolic umbilic catastrophe in its three-dimensional control space. The intersection of two separate sheets can be clearly seen, each sheet having on it the transformation of a cusp to a smooth boundary

critical points are in fact absolutely coincident, and we shall give an example of this procedure later in the chapter.

4.2 Quantum mechanical and Newtonian foundations

Before presenting our specific crystal analysis, we discuss first the appropriate general framework in which it should be viewed. A suitably general starting point for a discussion of the mechanical properties of many solids is provided by the steady-state quantum mechanical theory. In this we can eliminate the electrons from consideration by means of the Born–Oppenheimer adiabatic approximation[245] to obtain the Schrödinger equation for the nuclei alone, and we can include in the effective nuclear potential the energy of any applied conservative forces whose mechanical influence we wish to examine.

The solutions of the quantum mechanical problem are intimately related to the solutions of the corresponding Newtonian problem. Within the harmonic approximation, for example, the quantum energy levels are related in a simple fashion to the classical vibration frequencies, so many Newtonian results can be readily quantized, as we see, for example, in the work of Dean.[246] In particular, it is clear that a steady-state quantum mechanical solution can only be expected in the vicinity of a stable Newtonian equilibrium state. A logical first step towards a quantum mechanical solution is thus to make a static Newtonian stability analysis, and indeed this step will often supply all the information needed for a

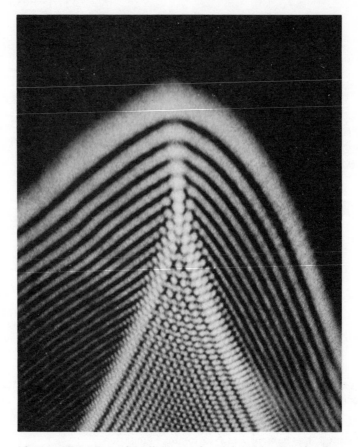

Photograph 7   Part of the light caustic produced by the refraction
of laser light by irregular 'bathroom window' glass. The cusped
inner curve and smooth outer curve make up a section through the
hyperbolic umbilic catastrophe. Because of the wave nature of light,
the caustic is decorated with a characteristic interference pattern.
Reproduced with the permission of M. V. Berry

bulk mechanical description of the solid. The work of Macmillan and Kelly is, for
example, typical of a continuing study of the mechanical properties of perfect
crystals within a static Newtonian approximation for the atoms using semi-
empirical interatomic potentials of the Lennard–Jones and Born–Mayer types.

The appropriate tool for this Newtonian analysis is the general theory of elastic
stability for discrete conservative systems,[36] and in line with this we can write the
effective potential V as a single-valued function of n generalized coordinates Q_i
defining admissible positions of the atoms and a loading parameter Λ defining,
for example, the overall stress level. We are thus concerned with the equilibrium
and stability of a discrete conservative mechanical system described by a total
potential energy function $V(Q_i, \Lambda)$, and our earlier work[36] offers established

perturbation procedures for examining the non-linear features of any branching points.

When the two critical points of our perfect mechanically stressed lattice are essentially distinct, explicit closed-form results for our unstable-symmetric branching point are readily applied, and we have already drawn upon their qualitative nature in our outline of Section 4.1. In the crystals field, however, the two critical points are as we have indicated often rather close (and can indeed be made totally coincident by the introduction of a second direct stress), and as we have explained it is then advantageous to use a general theory for the compound instability of a hill-top branching point.

Such a theory was not specifically developed in our earlier monograph, and since we wish to make use of it in our crystal fracture analysis, we shall present it briefly now.

4.3 Perturbation theory for hill-top branching

The general theory of this section is rather technical, and for this reason many readers may like to skip section 4.3 altogether, the necessary qualitative results having been already sketched. In doing this, readers should therefore have no difficulty in following the subsequent sections.

We consider then a discrete conservative system with the energy function $V(Q_i, \Lambda)$ and it will be sufficient for our purpose to suppose that we have only two (active) coordinates Q_1 and Q_2. Taking V to be symmetric in Q_1 the equilibrium equations $V_i = 0$ are assumed to give equilibrium paths with the form of those shown in Figure 62, and to study the hill-top branch in detail we introduce the new incremental coordinates

$$u_1 = Q_1, \qquad u_2 = Q_2 - Q_2^C. \tag{1}$$

In terms of these fixed coordinates we have the energy function

$$D(u_i, \Lambda) \equiv V(u_1, Q_2^C + u_2, \Lambda). \tag{2}$$

To establish the local form of the secondary equilibrium path we now write it in the parametric form

$$u_2 = u_2(u_1), \qquad \Lambda = \Lambda(u_1), \tag{3}$$

and we can substitute these expressions into the equilibrium equations $D_i = 0$ to give the identities

$$D_1\{u_1, u_2(u_1), \Lambda(u_1)\} \equiv 0, \qquad D_2\{u_1, u_2(u_1), \Lambda(u_1)\} \equiv 0. \tag{4}$$

We next generate a perturbation scheme to determine derivatives of $u_2(u_1)$ and $\Lambda(u_1)$ by differentiating these identities repeatedly with respect to u_1 to obtain

$$D_{11} + D_{12}u_2^{(1)} + D_1'\Lambda^{(1)} = 0, \tag{5}$$

$$D_{21} + D_{22}u_2^{(1)} + D_2'\Lambda^{(1)} = 0, \tag{6}$$

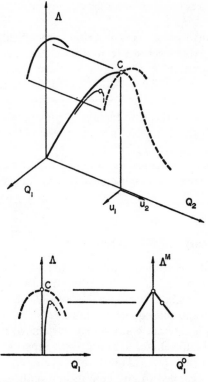

Figure 62 A hill-top branching point on the primary stress–strain curve of a crystal

$$(D_{111} + D_{112}u_2^{(1)} + D'_{11}\Lambda^{(1)}) + (D_{121} + D_{122}u_2^{(1)} + D'_{12}\Lambda^{(1)})u_2^{(1)} + D_{12}u_2^{(2)}$$
$$+ (D'_{11} + D'_{12}u_2^{(1)} + D''_1\Lambda^{(1)})\Lambda^{(1)} + D'_1\Lambda^{(2)} = 0, \quad (7)$$

$$(D_{211} + D_{212}u_2^{(1)} + D'_{21}\Lambda^{(1)}) + (D_{221} + D_{222}u_2^{(1)} + D'_{22}\Lambda^{(1)})u_2^{(1)} + D_{22}u_2^{(2)}$$
$$+ (D'_{21} + D'_{22}u_2^{(1)} + D''_2\Lambda^{(1)})\Lambda^{(1)} + D'_2\Lambda^{(2)} = 0, \quad (8)$$

etc., where a bracketed superscript indicates the number of differentiations with respect to u_1 and a prime denotes differentiation with respect to Λ.

We now evaluate these equations at the critical equilibrium state C, noting that because of the assumed symmetry of the energy function we have

$$D_{12}^C = D'^C_1 = 0, \qquad D_{122}^C = D''^C_1 = D'^C_{12} = 0, \qquad D_{111}^C = 0. \quad (9)$$

In particular, D_{ij}^C is diagonal, the coordinates u_i being thus *principal coordinates*, and because we are postulating a compound critical equilibrium state we have the two zero *stability coefficients*

$$D_{11}^C = 0, \qquad D_{22}^C = 0. \quad (10)$$

With these zero values, (5) gives on evaluation no information. For a limit point in the (Λ, Q_2) plane the general theory shows that we must have[36]

$$D_2'^C \neq 0, \tag{11}$$

and with this assumption, (6) yields the first derivative of interest,

$$\Lambda^{(1)C} = 0. \tag{12}$$

In the situation envisaged we have

$$D_{112}^C \neq 0, \tag{13}$$

so (7) gives us the second derivative of interest,

$$u_2^{(1)C} = 0, \tag{14}$$

and finally (8) yields

$$D_{211} + D_2'\Lambda^{(2)}|^C = 0 \tag{15}$$

so that

$$\Lambda^{(2)C} = -\left.\frac{D_{112}}{D_2'}\right|^C, \tag{16}$$

which is an expression for the initial curvature of the secondary equilibrium path at the branching point, corresponding to the first non-zero term of the Taylor series expansion of $\Lambda(u_1)$ about $u_1 = 0$.

So far we have only considered a 'perfect' system, and we now identify a corresponding imperfect system by the energy function $D(u_i, \Lambda, \varepsilon)$ where ε is an imperfection parameter, the vanishing of which will serve to retrieve the perfect system of our earlier consideration. The equilibrium solutions for such a system with $\varepsilon \neq 0$ will have the form of the light lines in Figure 63 and we are primarily interested in the maxima of these paths for which we write

$$\Lambda = \Lambda^M, \qquad u_i = u_i^M, \qquad \varepsilon = \varepsilon^M. \tag{17}$$

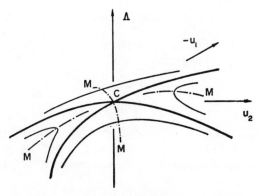

Figure 63 The hill-top branching point for perfect and imperfect systems

We would like to construct a series solution for $\Lambda^M(\varepsilon^M)$ along the trace of maxima (MM in Figure 63) and to do this we write this locus in the parametric form

$$\Lambda = \Lambda^M(u_1^M), \qquad u_2 = u_2^M(u_1^M), \qquad \varepsilon = \varepsilon^M(u_1^M). \tag{18}$$

The maxima are firstly equilibrium solutions and secondly critical equilibrium solutions, so clearly they must satisfy $D_i = 0$ and $\Delta = |D_{ij}| = 0$, and characterizing identities are written

$$D_1\{u_1^M, u_2^M(u_1^M), \Lambda^M(u_1^M), \varepsilon^M(u_1^M)\} \equiv 0, \tag{19}$$

$$D_2\{u_1^M, u_2^M(u_1^M), \Lambda^M(u_1^M), \varepsilon^M(u_1^M)\} \equiv 0, \tag{20}$$

$$\Delta\{u_1^M, u_2^M(u_1^M), \Lambda^M(u_1^M), \varepsilon^M(u_1^M)\} \equiv D_{11}D_{22} - D_{12}^2 \equiv 0. \tag{21}$$

Repeated differentiation of these identities with respect to u_1^M and evaluation at the critical point of the perfect system,

$$u_i^M = 0, \qquad \Lambda^M = \Lambda^C, \qquad \varepsilon^M = 0, \tag{22}$$

gives after some algebra the required derivatives

$$\varepsilon^{M(1)C} = 0, \tag{23}$$

$$\Lambda^{M(1)C} = 0, \tag{24}$$

$$\varepsilon^{M(2)C} = \mp 2\left(\frac{D_{112}}{\dot{D}_1}\right)\left(\frac{D_{112}}{D_{222}}\right)^{1/2}\Bigg|^C, \tag{25}$$

$$\Lambda^{M(2)C} = -\frac{2D_{112}}{D_2'}\Bigg|^C. \tag{26}$$

Here we have assumed, as is the case in our later application, that

$$\dot{D}_1^C \neq 0, \qquad \dot{D}_2^C = 0, \tag{27}$$

where a superimposed dot denotes differentiation with respect to ε. It is instructive to note that

$$\Lambda^{M(2)C} = 2\Lambda^{(2)C}. \tag{28}$$

If we now consider the Taylor series for $\Lambda^M(u_1^M)$ and $\varepsilon^M(u_1^M)$ about $u_1^M = 0$ and retain only the first relevant terms, then

$$\Lambda^M = \Lambda^C + \tfrac{1}{2}\Lambda^{M(2)C}(u_1^M)^2, \tag{29}$$

$$\varepsilon^M = \tfrac{1}{2}\varepsilon^{M(2)C}(u_1^M)^2, \tag{30}$$

where the leading derivatives $\Lambda^{M(2)C}$ and $\varepsilon^{M(2)C}$ have been determined by our perturbation scheme. Eliminating u_1^M between (29) and (30) gives the required local form of the imperfection sensitivity relationship $\Lambda^M(\varepsilon^M)$ as

$$\Lambda^M - \Lambda^C = \pm\left(\frac{\dot{D}_1^C}{D_2'^C}\right)\left(\frac{D_{222}^C}{D_{112}^C}\right)^{1/2}\varepsilon^M. \tag{31}$$

The details of this compound critical point are now apparent and are shown in Figure 60 and 63. The main difference from the distinct branching point is that the $\Lambda^M(\varepsilon^M)$ imperfection sensitivity is now less severe since the two-thirds power-law cusp has been replaced by a locally linear relationship.

Now in the crystals field we often encounter a branching point which is *strictly distinct* but which is nevertheless *quite close* to the maximum of the primary stress–strain curve. In this circumstance the general results with the two-thirds power law will apply very close to the branching point, so that there will indeed be a sharp cusp. Once we are some distance from the two critical points, however, it will seem as if they are coincident and we can expect the linear law to be more valid.

We shall indeed see in our application that although the critical points are strictly distinct, an *approximate compacted analysis* in which they are considered *strictly coincident* gives extremely good agreement with a numerical solution for the whole range of practical interest.

We might finally observe that the compound theory is actually easier to apply since it requires energy derivatives of lower order than those in the distinct theory. The former does, for example, only require the third-order partial derivatives D_{222}^C and D_{112}^C while the distinct branching point involves the fourth-order partial derivative D_{1111}^C.

4.4 Imperfection sensitivity in crystal fracture

Starting with a close-packed crystal of N identical atoms, we have initially $3N$ degrees of freedom corresponding to the coordinates X_i of the atoms. We study first, however, the response of a close-packed crystalline *sheet* of atoms under uniaxial tensile stress σ_{11} perpendicular to a close-packed direction. A planar, kinematically admissible homogeneous displacement field with only two degrees of freedom is employed, and a bifurcation point is located on the primary equilibrium path *just before* the limiting maximum.

The eigenvector is associated with a shearing strain ε_{12} and a small shearing stress σ_{12} is introduced as a corresponding 'imperfection'. The non-linear equilibrium and stability problem is then solved, firstly in an ad hoc numerical solution and secondly with the use of our general theory for hill-top branching. The two solutions are in excellent agreement, confirming the value of our concept of a *compacted* analysis.

We shall finally indicate how our planar and homogeneous displacement field for the single close-packed sheet of atoms can be used to generate an admissible homogeneous field for a face-centred cubic crystal or an admissible but strictly non-homogeneous field for a close-packed hexagonal crystal.

4.5 The Lennard–Jones potential for an atomic bond

We shall assume that the potential energy of interaction between any two atoms is some function $v(s)$, where s is the distance between centres, so that we have only spherically symmetric central-force two-body potentials. Considering further only *nearest-neighbour* interactions there will be no interatomic forces in the unloaded crystal which can thus be viewed as a close-packed array of spheres of diameter d where

$$\frac{dv}{ds}\bigg|_{s=d} = 0. \tag{32}$$

In particular, we shall take the Lennard–Jones potential used by Macmillan and Kelly[244] for argon in which

$$v(s) = A(Bs^{-12} - s^{-6}), \tag{33}$$

where A and B are physical constants. Using (32) we find immediately that

$$B = \tfrac{1}{2}d^6, \tag{34}$$

and writing $s = \gamma d$ we obtain

$$v(\gamma) = Ad^{-6}(\tfrac{1}{2}\gamma^{-12} - \gamma^{-6}), \tag{35}$$

which is the relation shown in Figure 64. The interatomic tensile force T is given by

$$T = \frac{dv}{ds} = \left(\frac{6A}{\gamma^7 d^7}\right)(1 - \gamma^{-6}) \tag{36}$$

and this relation (shown in Figure 64) attains its maximum value given by

$$\frac{Td^7}{6A} = \left(\frac{6}{13}\right)\left(\frac{7}{13}\right)^{7/6} \tag{37}$$

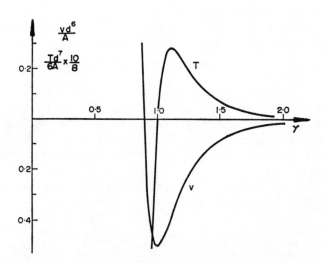

Figure 64 The interatomic potential V and the cohesive force T for two adjacent atoms

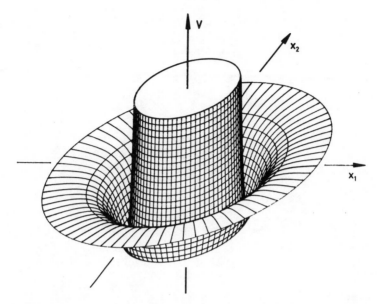

Figure 65 A computer drawing of the interaction potential for two atoms in a plane

for

$$\gamma = \left(\frac{13}{7}\right)^{1/6}. \tag{38}$$

Derivatives of v with respect to γ which will be required later can finally be written down as

$$v_\gamma = 6Ad^{-6}(-\gamma^{-13} + \gamma^{-7}), \qquad v_{\gamma\gamma} = 6Ad^{-6}(13\gamma^{-14} - 7\gamma^{-8}), \tag{39}$$

etc.

A computer drawing of our rotationally symmetric potential for a *plane* is shown in Figure 65, due to Richard Thompson.

4.6 The close-packed sheet

In our close-packed crystal we erect a set of three rectangular axes Ox_i oriented so that Ox_1 and Ox_2 lie in a close-packed plane of centres of atoms with Ox_2 lying along a close-packed column as shown in Figure 66. Here it will be seen that we have chosen to locate the coordinate origin mid-way between centres of atoms.

We consider next a displacement field with components $u_i(x_j)$ so that an atom originally at x_i will after deformation have the coordinates

$$X_i = x_i + u_i(x_j) \tag{40}$$

with, as elsewhere, use of the usual abbreviated notation for subscripts. If we

94

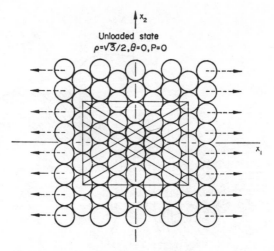

$\uparrow x_2$

Unloaded state
$\rho = \sqrt{3}/2, \theta = 0, P = 0$

x_1

Figure 66   The close-packed sheet of atoms in the
unloaded state

restrict attention to homogeneous deformations of the crystal, we can now write

$$u_i = c_{ij}x_j, \tag{41}$$

so that

$$X_i = x_i + c_{ij}x_j. \tag{42}$$

Choosing now the same finite strain definition as Macmillan and Kelly[244] we
can write

$$\varepsilon_{ij} = \tfrac{1}{2}(c_{ij} + c_{ji} + c_{ri}c_{rj}), \tag{43}$$

but as they observed it is convenient in calculations of this type to retain the c_{ij}'s
themselves as our deformation measures.

With our present restriction to homogeneous deformations we now have only
nine degrees of freedom for our crystal which we associate with the c_{ij} coefficients.
Of these nine degrees of freedom three will be associated with rigid-body
rotations and will be eliminated by suitable constraints.

We propose to load the crystal by a uniaxial tensile stress σ_{11} along the x_1 axis,
and we shall constrain the atoms in the $x_3 = 0$ plane to remain in that plane so
that we must set

$$c_{31} = c_{32} = 0. \tag{44}$$

Secondly, atoms lying on the x_1 axis will be constrained to remain on that axis so
that

$$c_{21} = 0. \tag{45}$$

We are left now with six degrees of freedom corresponding to the six non-zero c_{ij}
coefficients.

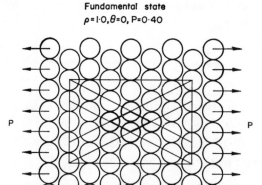

Fundamental state
$\rho = 1.0, \theta = 0, P = 0.40$

Figure 67 The close-packed sheet of atoms in the
fundamental state with $\rho = 1.0$

Considering the close-packed plane $x_3 = 0$, the applied stress σ_{11} will tend to
separate the close-packed columns as shown in Figure 67. It would then seem
plausible that at large direct strain ε_{11} these columns would become rotationally
unstable so that a shearing strain ε_{12} would develop as shown in Figure 68.

To explore this possibility firstly for just a single sheet of atoms we have the
three non-zero coefficients c_{11}, c_{12}, and c_{22} but we shall allow the plane only two
degrees of freedom by assuming (perhaps somewhat arbitrarily) that the close-
packed columns experience no change of length. The deformation can then be
described conveniently by the resolved distance between the columns which we
write as ρd and rotation of the columns θ as shown in Figure 68. The planar

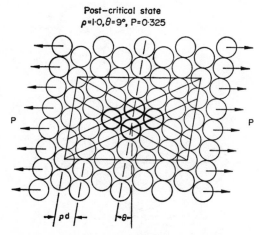

Post−critical state
$\rho = 1.0, \theta = 9°, P = 0.325$

Figure 68 The close-packed sheet of atoms
in the post-critical state with $\rho = 1.0$ and $\theta = 9°$

displacement coefficients can then be written in terms of our two generalized coordinates θ and ρ as

$$c_{11} = \left(\frac{2}{\sqrt{3}}\right)\rho - 1, \qquad c_{12} = \sin\theta, \qquad c_{22} = \cos\theta - 1, \tag{46}$$

with $c_{21} = 0$ as already specified.

The sheet is loaded primarily by the direct stress σ_{11}, and as an 'imperfection' we apply additionally a small shearing stress σ_{12}.

4.7 The four-atom unit cell

It is easy to see that with our assumptions the total potential energy of the infinite sheet is simply a multiple of that of the unit cell (of four atoms) shown in Figure 69. It will therefore be useful to focus our attention on this unit cell so that our system is precisely defined.

With our consideration of only *nearest-neighbour interactions* the unloaded unit cell rests as shown in the top diagram of Figure 69 with no interatomic forces. The two central atoms are assumed to maintain their unloaded distance d and the two independent generalized coordinates θ and ρ serve to define the deformation of the unit cell. The two convenient dependent variables α and β of Figure 69 can be expressed in terms of θ and ρ as

$$\alpha^2 = \tfrac{1}{4} + \rho^2 - \rho\sin\theta, \qquad \beta^2 = \tfrac{1}{4} + \rho^2 + \rho\sin\theta. \tag{47}$$

The unit cell is loaded by the direct axial forces F, which are related in a simple fashion to σ_{11} and by the shearing forces L which are related in a simple fashion to

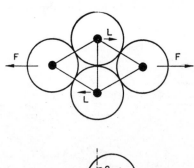

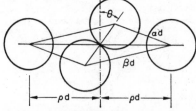

Figure 69 The four-atom unit cell in the undeformed and deformed states

σ_{12}. The potential energy of these forces is

$$-2F\rho d - Ld \sin \theta \qquad (48)$$

which is approximated here by

$$-2F\rho d - Ld\theta \qquad (49)$$

so the total potential energy of the unit cell can be written as

$$V(\theta, \rho, F, M) = 2v\{\alpha(\theta, \rho)\} + 2v\{\beta(\theta, \rho)\} - 2F\rho d - M\theta. \qquad (50)$$

Here we have written $Ld = M$ to emphasize the fact that our energy expression is *exact* if we care to think of the central atoms as loaded by a *moment M* rather than by the *forces L*.

Derivatives of V can now be written down as

$$V_\theta = 2v_\alpha \alpha_\theta + 2v_\beta \beta_\theta - M, \qquad V_\rho = 2v_\alpha \alpha_\rho + 2v_\beta \beta_\rho - 2Fd, \qquad (51)$$

etc., where

$$\alpha_\rho \equiv \frac{\partial \alpha}{\partial \rho} = \frac{2\rho - \sin \theta}{2\alpha}, \qquad \alpha_\theta \equiv \frac{\partial \alpha}{\partial \theta} = \frac{-\rho \cos \theta}{2\alpha}, \qquad (52)$$

etc.

For the 'perfect' system for which $M = 0$ the equilibrium equations

$$V_\theta = 0, \qquad V_\rho = 0 \qquad (53)$$

yield a primary (or fundamental) solution with $\theta = 0$ given by

$$4v_\gamma \gamma_\rho = 2Fd, \qquad (54)$$

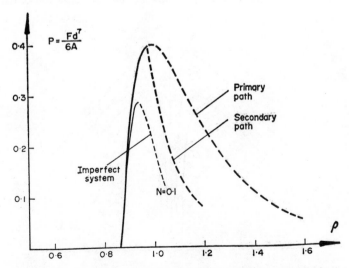

Figure 70 The numerical solutions for the infinite sheet of atoms showing the response of the perfect system and one imperfect system

where γ denotes either α or β, for which

$$P \equiv \frac{Fd^7}{6A} = \gamma^{-8}(1-\gamma^{-6})(4\gamma^2 - 1)^{1/2}. \tag{55}$$

In terms of ρ, equation (55) becomes

$$P = 2\rho\{(\rho^2 + \tfrac{1}{4})^{-4} - (\rho^2 + \tfrac{1}{4})^{-7}\}, \tag{56}$$

which is the relation shown in Figure 70.

Now on this primary path the symmetry of the system ensures that the derivative $V_{\theta\rho}$ is zero, so that the stability of the path is determined by the two *stability coefficients* $V_{\theta\theta}^F$ and $V_{\rho\rho}^F$ where F denotes evaluation on the path. The vanishing of $V_{\rho\rho}^F$ simply corresponds to the maximum of the path while the vanishing of $V_{\theta\theta}^F$ locates a point of bifurcation at which a θ-type deformation can develop.

Then, setting

$$V_{\theta\theta}^F = 0, \tag{57}$$

we find the critical value of γ,

$$\gamma^C = \left(\frac{7}{4}\right)^{1/6}, \tag{58}$$

and we observe that the bifurcation occurs just before the limiting maximum of the primary path as shown in Figure 70.

Solving the equilibrium equations numerically for the 'perfect' system with no shearing stress and for 'imperfect' systems with $N \equiv Md^6/6A \neq 0$ we have obtained the curves given in Figure 70 and 71. The deformation of the sheet of

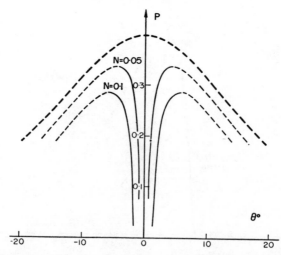

Figure 71 The unstable-symmetric branching point on a plot of the direct stress parameter P against the shearing strain parameter θ

atoms is drawn to scale for one of the predicted post-critical equilibrium states in Figure 68.

4.8 Application of the general theory

Identifying Λ with P, u_1 with θ, and ε with N, the asymptotic results of our extended general theory can be written in the terms of our crystallographic study as

$$\frac{d^2 P}{d\theta^2}\bigg|^C = -\frac{V_{\theta\theta\rho}}{V'_\rho}\bigg|^C , \tag{59}$$

$$P^M - P^C = \pm \left(\frac{\dot{V}_\theta}{V'_\rho}\right)\left(\frac{V_{\rho\rho\rho}}{V_{\theta\theta\rho}}\right)^{1/2} N^M \bigg|^C . \tag{60}$$

Equation (59) gives an expression for the initial curvature of the secondary equilibrium path, while (60) gives a failure-stress locus on a graph of P^M versus N^M. Equations (59) and (60) apply strictly to a compound critical point at which the energy derivatives should be evaluated, but it is our intention to compact the analysis and to proceed as if there were such a compound point. To do this, we can simply adopt these general expressions and evaluate the derivatives either at the *branching point* or at the *limit point*.

We have chosen to make the evaluations at the real branching point denoted by C, and the required values can be listed as

$$\frac{\partial^3 V}{\partial\theta^2 \partial\rho}\bigg|^C \equiv V^C_{\theta\theta\rho} = -\frac{87 \cdot 78 A}{d^6},$$

$$\frac{\partial^2 V}{\partial\rho\partial P}\bigg|^C \equiv V'^C_\rho = -\frac{12 A}{d^6},$$

$$\frac{\partial^2 V}{\partial\theta\partial N}\bigg|^C \equiv \dot{V}^C_\theta = -\frac{6 A}{d^6}, \tag{61}$$

$$\frac{\partial^3 V}{\partial\rho^3}\bigg|^C \equiv V^C_{\rho\rho\rho} = -\frac{351 \cdot 1 A}{d^6}.$$

Using (61) we obtain immediately the required solutions which are compared with the results of the numerical analysis in Figure 72 and 73. The agreement is seen to be exceptionally good.

4.9 The three-dimensional lattice

If we care to make the rather drastic assumption that all bonds between the close-packed planes $x_3 = $ constant retain their original length d we find that we can stack identical deformed sheets to generate kinematically admissible displace-

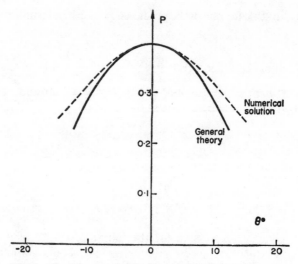

Figure 72   The post-critical response of the perfect system as predicted by the *compacted* general theory and the numerical solution

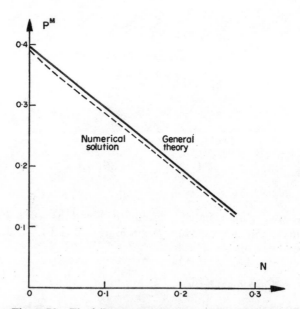

Figure 73   The failure-stress locus on a plot of the direct stress against the shearing stress as predicted by the *compacted* general theory and the numerical solution

ment fields for a three-dimensional lattice. In the case of a face-centred cubic crystal the displacement field will be *homogeneous*, but in the case of a close-packed hexagonal crystal the field will be strictly *non-homogeneous*.

For the face-centred cubic crystal the displacement coefficients would be

$$
c_{ij} = \begin{bmatrix} 2\rho \left/ \sqrt{3}-1 \right. & \sin\theta & \sqrt{\tfrac{3}{2}}(\rho^2 - \tfrac{3}{4})/6\rho \\ 0 & \cos\theta - 1 & -\sqrt{\tfrac{3}{2}}\tan\theta(\rho^2 - \tfrac{1}{4})/2\rho \\ 0 & 0 & \tfrac{1}{2}\sqrt{\tfrac{3}{2}}\{3 - (\rho^2 - \tfrac{1}{4})^2/\rho^2\cos^2\theta\}^{1/2} - 1 \end{bmatrix}, \quad (62)
$$

so that on the primary path we have

$$
c_{ij}^{\mathrm{F}} = \begin{bmatrix} 2\rho \left/ \sqrt{3}-1 \right. & 0 & \sqrt{\tfrac{3}{2}}(\rho^2 - \tfrac{3}{4})/6\rho \\ 0 & 0 & 0 \\ 0 & 0 & \tfrac{1}{2}\sqrt{\tfrac{3}{2}}\{3 - (\rho^2 - \tfrac{1}{4})^2/\rho^2\}^{1/2} - 1 \end{bmatrix}, \quad (63)
$$

while the eigenvector of our distinct branching point is given by

$$
\left.\frac{\partial c_{ij}}{\partial\theta}\right|^{\mathrm{C}} = \begin{bmatrix} 0 & 1 & 0 \\ 0 & 0 & -\sqrt{\tfrac{3}{2}}(\rho^2 - \tfrac{1}{4})/2\rho|^{\mathrm{C}} \\ 0 & 0 & 0 \end{bmatrix} \quad (64)
$$

4.10 Further work

In a continuation of this analysis, we have eliminated the approximation in passing from equation (48) to equation (49) and have relaxed the lateral constraint to allow for full Poisson contraction: we have also studied the related problem in which c_{21} is allowed to be non-zero. In all these cases, the numerical results are very close to those we have presented.

We have also[83] included the effect of a direct lateral stress σ_{22} and calculated the critical value of this stress for the generation of a *true* hill-top branching point. A perturbation analysis about this compound critical state would give the hyperbolic umbilic surface of Figure 61, as we have discussed.

Excellent discussions of the stability of ideal crystals are given by Hill[247] and Hill and Milstein.[248] In the former, Hill emphasizes the need, in calculations of the present type, to consider carefully the tractions applied to the homogeneous element, since the instability loads naturally depend on their precise definition. In the work reported here, we have prescribed dead loads and to avoid a bifurcation in the unloaded state we have added what seems to be an intelligent rotational constraint. As an alternative to dead loads one could consider various types of follower stresses, although it is not clear how these could be realized in an experimental test. A long-term aim is clearly to use the calculated strengths to predict failure in a position of stress concentration in a loaded body, but since the elastic environment in such a body will vary from problem to problem, it is far from clear which is the most realistic loading to use for our homogeneous calculations.

The conclusions of our analysis could be highly relevant to fracture mechanics studies, since an unstable bifurcation in the tensile zone at the tip of a crack could be a mechanism for destroying the symmetry of a plane propagating crack. Because we have here restricted attention to homogeneous deformations we have employed a small symmetry-breaking shearing stress as an imperfection. It is clear, however, that a dislocation would be equally valid as a trigger for the imperfection sensitivity, which can thus be seen as an asymptotic manifestation of the weakening action of a dislocation.

Rather analogous work on the stability of molecules in chemistry is outlined in a most readable article by Coulson,[249] while the atomic foundations of continuum mechanics are discussed by Mindlin.[250] Recent work on bifurcation phenomena in the plane tension test[251] and shear band formation[243] are important related studies.

CHAPTER 5

Spontaneous Order in Biochemical Reactions and Developmental Biology

Bifurcation decreases entropy. *Helen Petard*

As so neatly summarized by Helen Petard, bifurcations are key factors in the spontaneous emergence of patterns of spatiotemporal organization in chemical and biochemical systems. Indeed, it is clear *a priori* that if an initially homogeneous state is to organize itself in space or in time, then there must be some form of symmetry-breaking instability. Such self-organizing phenomena are fundamental to the understanding of morphogenesis in biological systems,

Figure 74 Spiral waves in the Belousov–Zhabotinski reaction. This remarkable picture of spatial order in an initially homogeneous chemical mixture is due to Winfree and we would refer readers to his excellent discussion of the reaction.[101] It is reproduced from A. T. Winfree, Spatial and temporal organization in the Zhabotinski reaction, *Advances in Biological and Medical Physics*, **16**, 115 (1977), with the permission of Academic Press

104

such as the development of differentiated bone and muscle in a growing organism.

This creation of order can, by reason of the second law of thermodynamics, only occur in an *open* system, which moreover must be essentially non-linear in behaviour. For such a system a self-organizing process is accompanied by an instability of a path of steady states, the *thermodynamic branch*, showing equilibrium-like behaviour. Beyond this instability an initially homogeneous system has the possibility of attaining an ordered configuration or *dissipative structure*.[252] Experimental evidence for the formation of such dissipative structures is available for both chemical and biochemical reactions.

A classical example of a self-organizing chemical reaction is the Belousov–Zhabotinski reaction. Here a shaken homogeneous chemical mixture, if left in a shallow dish, can organize itself into spiral patterns like those shown in Figure 74. Because such a dish is essentially a closed system, this self-organization is here only temporary and eventually the system reverts to a homogeneous chaotic state: the chemical 'organism' dies! A permanently organized state can, however, be maintained if appropriate chemicals are fed continuously into and out of such a system. Similar organization in a test-tube is shown in Photo 8.

5.1 The Brusselator model chemical reaction

Much theoretical interest has focused on a trimolecular *model* system, the so-called *Brusselator*, which is one of the simplest to exhibit this pheno-

Photograph 8 Spontaneous emergence of spatial patterns in the Zhabotinski reaction.[233] Equal volumes of $Ce_2(SO_4)_3$, $KBrO_3$, $CH_2(COOH)_2$, and H_2SO_4, together with a little redox indicator were stirred thoroughly. A quantity of the *homogeneous* mixture was poured into a test-tube kept at a constant 21°C. *Temporal* oscillations were immediately observed, the colour changing periodically from red, indicating an excess of Ce^{3+}, to blue, showing an excess of Ce^{4+}, with a period of approximately 4 minutes. After a number of such oscillations a small concentration inhomogeneity appeared, from which the alternate red and blue layers were formed. Because the test-tube is a closed system, this *spatial* structure can only be maintained for a limited time, here about 30 minutes, after which the system approaches equilibrium and goes back to a homogeneous distribution of matter. Photograph by courtesy of M. Herschkowitz-Kaufman

menon.[100,233] The *real* Belousov–Zhabotinski reaction is by contrast an exceedingly complex reaction which is even now not fully understood.

This model considers the hypothetical reactions:

$$A \rightarrow X$$
$$B + X \rightarrow Y + D$$
$$2X + Y \rightarrow 3X$$
$$X \rightarrow E$$

the trimolecular step being seen in the third reaction. Here A, B, D, and E are initial and final products, whose concentrations are imagined to be imposed as constants throughout. All reaction steps are here assumed to be irreversible with rate constants equal to unity.

The reactions are shown diagramatically in Figure 75 and we see that the system is indeed open with chemicals A and B continuously entering and chemicals D and E constantly leaving the system.

Using the same letters to denote the *concentrations* of the chemicals, the rate of production of X in the first reaction is simply A, while the rate of loss of X in the second equation is the product BX. The net rate of production of X in the third trimolecular step is $X^2 Y$, and finally the rate of loss of X in the fourth reaction is X.

Thus we can write

$$\frac{\partial X}{\partial t} = A - (B + 1)X + X^2 Y$$

and similarly

$$\frac{\partial Y}{\partial t} = BX - X^2 Y.$$

These are the coupled non-linear rate equations that can be solved for the time evolution of X and Y with A and B as prescribed constants.

Setting the rates equal to zero gives us the primary solution of the

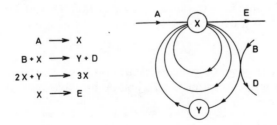

Figure 75 The Brusselator trimolecular model chemical reaction with two variable intermediates X and Y and the initial and final products A, B, D, and E. Each closed loop containing one intermediate implies the presence of one mole of this intermediate in the corresponding reaction step

thermodynamic branch

$$X = A, \qquad Y = \frac{B}{A}$$

and we write

$$X = A + x, \qquad Y = \frac{B}{A} + y$$

where x and y are now changes in concentration from the primary values. The evolution equations are now

$$\dot{x} = \underline{x(B-1) + y(A^2)} + x^2\frac{B}{A} + 2xyA + x^2y,$$

$$\dot{y} = \underline{x(-B) + y(-A^2)} - x^2\frac{B}{A} - 2xyA - x^2y,$$

and for a linear stability analysis we need retain only the underlined terms. The linear equations can therefore be written in the standard form

$$\dot{x} = c_{11}x + c_{12}y,$$
$$\dot{y} = c_{21}x + c_{22}y,$$

where

$$c_{11} = B - 1, \qquad c_{12} = A^2,$$
$$c_{21} = -B, \qquad c_{22} = -A^2,$$

and the characteristic equation becomes

$$\lambda^2 - \lambda(c_{11} + c_{22}) + c_{11}c_{22} - c_{12}c_{21} = 0.$$

Now the required coefficients are

$$c_{11}c_{22} - c_{12}c_{21} = A^2$$

and

$$-(c_{11} + c_{22}) = 1 + A^2 - B.$$

Since A is necessarily positive, we can never have a static instability at which the first coefficient would have to vanish, but we see that the vanishing of the second coefficient predicts a dynamic instability at

$$B^C = 1 + A^2.$$

Setting $A = 1$ and regarding the concentration B as a control parameter, we see that we have a stable focus for $B < B^C$ and an unstable focus for $B > B^C$ where

$$B^C = 2.$$

We have in fact at $B = B^C$ a dynamic Hopf bifurcation at which a stable super-critical limit cycle is generated, and some computed phase portraits from the full non-linear rate equations are shown in the following three figures.

Figure 76 shows a damped stable chemical oscillation for $B = 1.5$ with for reference the stable limit cycle generated at $B = 3$. At this sub-critical value of

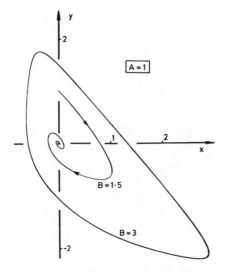

Figure 76 Damped oscillatory motion
to a stable focus at a sub-critical value of
B

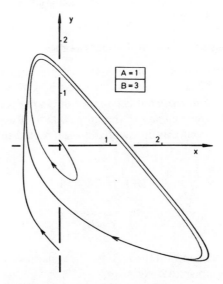

Figure 77 Motions to a stable limit cycle of sustained chemical oscillation at a
super-critical value of *B*

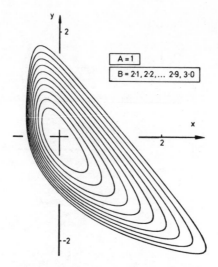

Figure 78 A nest of growing limit cycles generated as B increases from its critical value of two

$B = 1.5$, disturbances are damped out, and the chemical concentrations always return to those of the thermodynamic branch, $x = y = 0$.

Figure 77 shows the system moving to a stable super-critical limit cycle for $B = 3$, and here a permanent chemical oscillation is established with the well-defined periodic time of this cycle.

Figure 78 shows the nest of limit cycles, generated at $B^C = 2$, for a series of super-critical values of B.

5.2 Creation of spatial order in thermodynamics

What we are seeing here is the temporal organization of a spontaneously oscillating chemical reaction, and we proceed now to look at spatial organization. To do this we suppose our chemical model to be smeared out along a line with distance coordinate r lying in the range 0 to 1. If we allow *diffusion* of the chemicals X and Y along this line, with diffusion coefficients D_1 and D_2, the rate equations are now

$$\frac{\partial X}{\partial t} = A - (B+1)X + X^2 Y + D_1 \frac{\partial^2 X}{\partial r^2},$$

$$\frac{\partial Y}{\partial t} = BX - X^2 Y + D_2 \frac{\partial^2 Y}{\partial r^2}.$$

With A and B held constant throughout the system, we again have the primary solution involving no spatial variation

$$X = A, \qquad Y = \frac{B}{A}$$

which is compatible with the assumed boundary conditions

$$X(0,t) = X(1,t) = A,$$

$$Y(0,t) = Y(1,t) = \frac{B}{A}.$$

We can again perform a stability analysis in terms of the concentration changes x and y, and the necessary *linear* equations are now

$$\frac{\partial x}{\partial t} = (B-1)x + A^2 y + D_1 \frac{\partial^2 x}{\partial r^2},$$

$$\frac{\partial y}{\partial t} = -Bx - A^2 y + D_2 \frac{\partial^2 y}{\partial r^2}.$$

The solution of these partial differential equations can be written as

$$x(r,t) = x_0 e^{wt} \sin n\pi r,$$

$$y(r,t) = y_0 e^{wt} \sin n\pi r,$$

giving us both steady-state and time-periodic eigenvalue solutions.

We see that the system can now give sinusoidally varying *spatial patterns*, and a non-linear perturbation study due to Auchmuty and Nicolis[253] is summarized in Figure 79. Here if the spatial wave number n is even we have a stable-symmetric static bifurcation leading to stable super-critical dissipative structures, while if n is

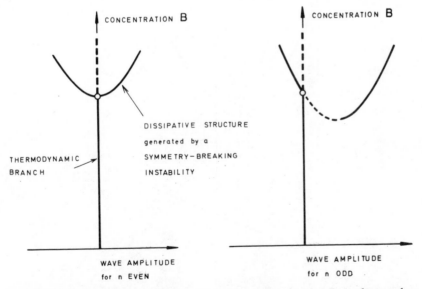

Figure 79 Spontaneous emergence of spatial order in the Brusselator when distributed along a line. Bifurcations involving steady-state solutions showing the creation of stable dissipative structures after the loss of stability of the primary thermodynamic solution

110

odd we have an asymmetric static bifuraction leading to trans-critical dissipative structures with the stable regions indicated by a solid line. Thus we see that with the chemicals distributed in space along a line both spatial and temporal patterns can be spontaneously generated by an instability of the primary thermodynamic branch.

Two-dimensional studies of this model chemical reaction have also been made,[254] and recent work by Mahar and Matkowsky in New York and Herschkowitz-Kaufman in Brussels has extended the present one-dimensional analysis to demonstrate the existence of secondary bifurcations.[100]

These secondary bifurcations can be associated with the proximity of a stable-symmetric and an asymmetric point of bifurcation, in which case they may be related to an underlying umbilic catastrophe or with the proximity of two stable-symmetric points of bifurcation, in which case they may be related to an underlying double-cusp catastrophe. The importance of these secondary bifurcations is that they may imply the stabilization of the path from the second thermodynamic eigenvalue, giving us an alternative stable dissipative structure as illustrated in Figure 80.

Further accounts of these recent findings can be found in the authoritative monograph of Nicolis and Prigogine.[100] It is felt that the instability and

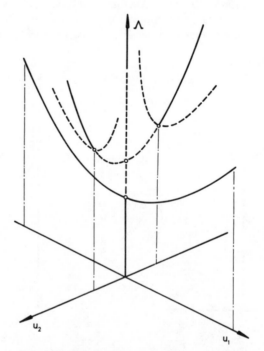

Figure 80 Restabilization at a secondary branching point. This phenomenon is of interest to thermodynamicists since it may give rise to alternative dissipative structures far from thermodynamic equilibrium

oscillation of chemical and biochemical systems may be relevant to the original emergence of life from the primeval soup and to the functioning of the mysterious biological clocks.[255]

5.3 Cell differentiation in developmental biology

We shall finally look briefly at Zeeman's analysis of cell differentiation, which models morphogenesis in biology at a quite different level. So as a simplified model in biology, we consider the differentiation of essentially identical cells into, say, bone and muscle during the development of an embryo. We regard the cells as a multitude of identical systems controlled by the local concentration of a number of chemicals.

We consider a one-dimensional region of an organism with the cells distributed along the axis Ox. The concentrations vary with x which we can thus take as a fundamental control parameter together with time, so that we have two controls Λ^1 and Λ^2.

With the four hypothesis of homeostasis, continuity, differentiation, and repeatability, and assuming that the biochemistry of a cell can be modelled by a gradient dynamic, Zeeman[10] established Figure 81 to represent the process of differentiation. Here the single internal state variable Q represents a continuous measure of the degree of differentiation of a cell.

Initially, Q varies continuously with x from the pre-muscle cells at m to the pre-bone cells at b, while at the end of our time interval when differentiation is complete there is a discontinuity in Q between the muscle at M and the bone at B. Between these two time sections, Zeeman draws a cusp catastrophe, carefully inclined so that it does not point precisely parallel to the time axis. This he considers an essential feature to ensure structural stability and hence repeatability, although an aligned cusp gives a simpler scheme which might be an acceptable approximation in certain circumstances.

Cells with x less than x_C develop in a continuous smooth fashion to give muscle at M. Cells with x between x_C and x_A, however, encounter a dynamic jump, associated perhaps with the sudden 'switching on of gene systems', during which there is a step change in Q towards the production of muscle. The timing of this jump varies with x, so that a *developmental wave* travels through the body, the jumps getting more severe until the wave is arrested at A. Here we have our asymmetric point of bifurcation dividing sub-critical cells with x less than x_A which develop discontinuously into muscle from super-critical cells with x greater than x_A that exhibit no instability and develop smoothly into bone.

The developmental wave is predicted to start with a finite velocity but slow down parabolically in a gradual fashion. The final result is the formation of a sharp spatial boundary between muscle and bone at x_A. The bifurcation at A is the essentially divergent feature, although the observable effects of our 'primary' wave may not appear for some time. This possible delay leads to the concept of secondary waves which are delayed manifestations of the primary. It is the secondary wave of physical manifestation that may well signal the release of chemical energy to provide the physical energy for morphogenesis, the creation

112

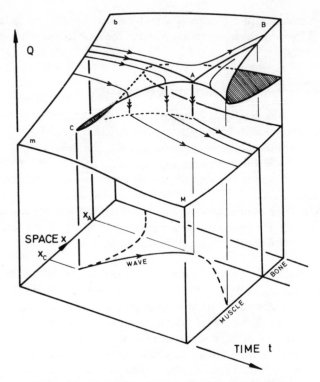

Figure 81 The differentiation of cells in developmental biology. The cells of an embryo are supposed spaced along a line with space coordinate x. Those with low x must develop into muscle while those with high x must develop into bone. The single internal variable Q can be thought of as representing the 'boniness' of a cell. The picture shows Zeeman's representation of the process as an inclined cusp catastrophe. Reproduced with the permission of the American Mathematical Society from E. C. Zeeman, Primary and secondary waves in developmental biology, *Lectures on Maths in the Life Sciences*, **7**, 69 (1974)

of spatial forms. It is interesting to notice the close similarity between Figure 81 and Benjamin's study of flow between rotating cylinders that we discuss in our chapter on hydrodynamics.

The above developmental ideas are applied in detail by Zeeman to the embryonic development of frogs and the culmination of slime mould. One further application of catastrophe theory in biology, also due to Zeeman, can be found in his work on the heartbeat.

In the large and developing field of biological morphogenesis Zeeman's model is an example of an ad hoc attack at the central problem at a high and abstract level of modelling. A more continuous and detailed study grounded in the specific underlying biochemistry is due to the Brussels school of I. Prigogine and G. Nicolis, as we have already indicated, and we would finally draw attention to the fine work of Erneux and Hiernaux.

In a recent paper[256] these authors emphasize that 'Pattern formation is a complex process whereby cells, possessing the same genetic potentialities, acquire various states of molecular differentiation, leading to a well-characterized spatial structure. The formation of those differentiated patterns has been related to the existence of inhomogeneous distribution of chemicals, the so-called morphogens, throughout the morphogenetic fields.'[257] They then proceed to show the special relevance of secondary bifurcations to the thermodynamics of these processes.

The same authors have also made a bifurcational and numerical analysis[258] of Turing's classic theory of morphogenesis,[259] while Nicolis and Malek-Mansour[260] have shown that in non-equilibrium phase transitions and chemical reactions a *master equation* approach gives different answers from the Maxwell construction employed by Thom.

We shall see self-organizational problems similar to those of this present chapter when we study the neural dynamics of the brain in the final chapter of the book.

CHAPTER 6

Population Dynamics and Evolution of a Prey–Predator Ecology

The dynamics, evolution, and organization of ecological systems have been intensively studied in recent years, as can be seen for instance in the final chapters of Nicolis and Prigogine,[100] and we present here some recent work[103] with the kind permission of the Institute of Mathematics and Its Applications.

The classical Lotka–Volterra equations form an interesting and instructive introduction to the population dynamics of a prey–predator ecosystem. Phase trajectories can be readily traced using an electronic pocket calculator, and a fox–rabbit board game is introduced here that realistically models the equations with an added random stochastic element. Interest in the game centres on the possible extinction or explosion of the rabbit population.

The differential equations governing the growth, decay, and general evolution of interacting biological species are similar in structure and form to those that we have encountered in chemical kinetics, and we shall see how the population dynamics of a simple prey–predator ecosystem can exhibit closed oscillations akin to the stable vibrations of an undamped pendulum.

6.1 Lotka–Volterra equations

Following Nicolis and Prigogine,[100] let us begin our mathematical modelling by supposing that in the presence of food, A, the members of a biological population, X, reproduce at a rate proportional to the product AX, so that

$$\frac{\mathrm{d}X}{\mathrm{d}t} = k_1 AX.$$

Here k_1 is a constant, and we use the X and A to represent both the population and food, and also a suitable measure of these quantities. In a similar way the rate of mortality can be assumed to be proportional to the current population level X through a pair of constants, k_3 and B, so that

$$\frac{\mathrm{d}X}{\mathrm{d}t} = -k_3 BX.$$

Combining these two equations, the resulting net rate of increase of the population is

$$\frac{dX}{dt} = (k_1 A - k_3 B)X.$$

If the food supply, A, were constant, this equation would predict a population explosion through an *exponential* growth of X, known as *Malthusian* growth, if $k_1 A > k_3 B$. Alternatively, if $k_1 A < k_3 B$, the rate equation predicts an exponential decay to zero population.

These unsatisfactory results lead to the conclusion that in animal populations the ecological system is stabilized by the limited availability of food, the rate of consumption of which must thus be included in the mathematical modelling.

To illustrate this we consider a simple model ecology involving a single prey, say a rabbit, with population X living in the presence of a single predator, say a fox, with population Y. We assume that the rabbits have an unlimited reservoir of food in the form of vegetable matter so that their reproduction follows the earlier law

$$\frac{dX}{dt} = k_1 AX,$$

where both k_1 and A are now constants, and that they die *only* as a result of an interaction with a fox. This latter condition presumes that there is always a sufficient number of foxes to ensure that essentially no rabbits die of old age. The predators, however, are assumed to die natural deaths according to the earlier law

$$\frac{dY}{dt} = -k_3 BY,$$

where k_3 and B are constants.

The *coupling* between the populations arises from the capture of a prey by a predator which serves to decrease the rabbit population and increase the fox population, since successful fox breeding is assumed to be proportional to their food supply. The probability of a fox catching a rabbit, and hence the overall rate of capture, might clearly be proportional to the *product* of the fox and rabbit populations, XY, so we add the interaction rates

$$\frac{dX}{dt} = -k_2 XY, \qquad \frac{dY}{dt} = +k_2 XY,$$

where k_2 is a constant.

Combining these distinct birth and death contributions we obtain the complete *non-linear coupled* evolution equations

$$\dot{X} = k_1 AX - k_2 XY,$$
$$\dot{Y} = k_2 XY - k_3 BY,$$

(1)

where a dot denotes differentiation with respect to time t. These are the classical

Lotka–Volterra dynamical equations describing a simple prey–predator ecology. They have been used for many years to model basic biological phenomena such as biological clocks and time-dependent neural networks.

We may notice in passing that these are precisely the rate equations that would arise from the set of hypothetical chemical reactions, with rates k_1, k_2 and k_3

$$A + X \xrightarrow{k_1} 2X,$$
$$X + Y \xrightarrow{k_2} 2Y,$$
$$Y + B \xrightarrow{k_3} E + B,$$

if the concentration of chemicals A and B were maintained constant and uniform inside the reaction vessel.

6.2 Stability of the steady state

By equating the rates of equation (1) to zero we obtain, apart from the trivial and uninteresting solution $X = 0$, $Y = 0$, the single steady-state solution

$$X_s = \frac{k_3 B}{k_2} \quad \text{and} \quad Y_s = \frac{k_1 A}{k_2}. \tag{2}$$

If the populations had initially these values, the numbers of prey and predators would remain constant in time according to this deterministic model. In a more realistic stochastic model, random fluctuations would have to be incorporated.

To examine the stability of this steady state we write

$$X = X_s + x,$$
$$Y = Y_s + y,$$

and substituting in equation (1) we assume that the increments x and y are small quantities so that their products can be ignored. We thus obtain the *linearized* equations

$$\dot{x} = -k_3 B y,$$
$$\dot{y} = +k_1 A x, \tag{3}$$

which describe small population changes around the steady state. Eliminating y between these two equations gives

$$\ddot{x} + k_1 k_3 A B x = 0, \tag{4}$$

which is the well-known equation of a simple harmonic oscillator of circular frequency

$$w = (k_1 k_3 A B)^{1/2} \tag{5}$$

and periodic time $T = 2\pi/w$.

The phase trajectories in X, Y space are thus concentric ellipses for small deviations from X_s, Y_s, and the steady state is thus stable. For larger finite oscillations about X_s, Y_s the phase trajectories are no longer elliptical, but they remain *closed* curves with, however, a continuous change in the periodic time. The large and small amplitude behaviour is thus entirely analogous to that of an *undamped* pendulum, and the detailed relationships between the *Lotka–Volterra* equations and the *Hamiltonian* equations of classical mechanics are highlighted by Nicolis and Prigogine in their recent monograph. They show that the constant of motion in the Lotka–Volterra equations is closely related to the excess entropy, $\delta^2 S$, around the reference state.

We should note carefully here that a linear prediction of elliptical centres does not in general guarantee centres in the corresponding non-linear system. Exclusively non-linear damping in a mechanical system could, for example, give asymptotically stable foci, even though the linearization, with no damping, predicted centres.

Because of the constantly varying periodic time, the dynamical motions along the phase trajectories are themselves only orbitally stable, and random stochastic fluctuations will induce a constant drifting between orbits in contrast to the convergence to a limit cycle in the Brusselator chemical model.

Three phase trajectories are shown in Figure 82 for which the constants have been set equal to unity,

$$k_1 A = k_3 B = k_2 = 1,$$

which gives the steady state

$$X_s = Y_s = 1,$$

and the linear theory ellipses become, in this case, circles.

The space X, Y is in reality filled with an infinity of such nesting phase

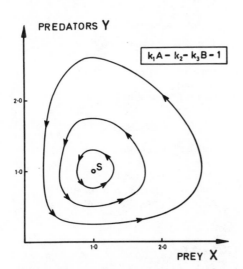

Figure 82 Closed phase trajectories for the prey–predator model, showing analogy with an undamped pendulum

118

trajectories, each with its own periodic time, and the trajectory through any starting point summarizes the dynamical evolution of the ecosystem. Thus if the system starts away from the steady state, S, it will exhibit undamped oscillations about X_s, Y_s, the amplitude of the oscillations being governed by the original departure from S.

6.3 Numerical solutions by finite differences

The differential equations (1) can be replaced for an approximate numerical calculation by the finite difference equations

$$\Delta X = (k_1 AX - k_2 XY)\Delta t,$$
$$\Delta Y = (k_2 XY - k_3 BY)\Delta t, \tag{6}$$

Table 3 A simple calculation sequence for the phase trajectories using a pocket electronic calculator

---

Parameters: $L = k_1 A/k_2$, $M = k_3 B/k_2$, $N = k_2 \Delta t$

$X_0 \Sigma 1$	set X_0 in store 1
$Y_0 S2$	set Y_0 in store 2
L	Read L
$-$	subtract
$R2$	recall Y_i from store 2
$\times$	multiply
$R1$	recall X_i from store 1
$\times$	multiply
N	read N
$=$	form ΔX
$\Sigma 1$	add ΔX to store 1
$R1$	Recall X_{i+1} from store 1
$-$	subtract
M	read M
$\times$	multiply
$R2$	recall Y_i from store 2
$\times$	multiply
N	read N
$+$	add
$R2$	recall Y_i from store 2
$=$	form $Y_i + \Delta Y$
$S2$	store $Y_i + \Delta Y$ in store 2
$R1$	recall and *print* X_{i+1} from store 1
$R2$	recall and *print* Y_{i+1} from store 2

Equations are:

$$(L - Y) XN = \Delta X,$$
$$(X - M) YN = \Delta Y.$$

Computer: Commodore SR4148R.

---

so if we are currently at a point $X = X_i$, $Y = Y_i$ we can take the new state after a small time increment Δt to be

$$X_{i+1} = X_i + \Delta X,$$
$$Y_{i+1} = Y_i + \Delta Y.$$

We can now recalculate ΔX and ΔY by substituting for state $i + 1$ in (6) and take a second time increment to yield

$$X_{i+2} = X_{i+1} + \Delta X,$$
$$Y_{i+2} = Y_{i+1} + \Delta Y,$$

etc. A sequence similar to this can be readily programmed for a pocket electronic calculator as shown in Table 3 for a Commodore SR4148R. Notice that this sequence differs slightly from the one outlined by the more convenient use of X_{i+1} in the calculation of Y_{i+1}.

The results of two such calculations for the previously employed constants of unity and with the time step of $\Delta t = \frac{1}{4}$ are shown in Figure 83. The outer trajectory was started with $i = 0$ at $X = 1$, $Y = 2.6$, and was traced to $i = 28$ where the phase trajectory is approximately closed. The 28 time steps show that the periodic time for an oscillation of this amplitude is approximately $T = 28/4 = 7$. The inner trajectory was started with $i = 0$ at $X = 1$, $Y = 1.75$ and was traced to an approximate closure at $i = 25$. The periodic time is here approximately 6.5 in

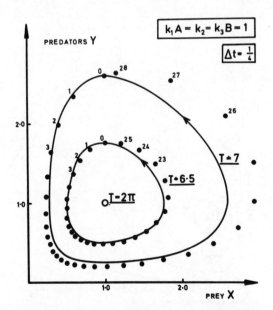

Figure 83 Simple estimation of the trajectories using a pocket electronic calculator. The constant time steps employed highlight the varying phase velocity

120

comparison with the outer time of 7 and the small-amplitude linear-theory time of 2π.

The varying phase velocity is nicely seen on this figure and we see that it increases with distance from the origin of coordinates, a feature that will be more pronounced in our later studies. The accuracy of the numerical solutions, which are compared here with the exact curves taken from Figure 82, is acceptable for our qualitative purposes and it can, of course, be increased by reducing the time step Δt.

To model a proposed evolution game that we shall outline in the following section, some further numerical calculations are shown in Figure 84 for the case of

$$k_1 A = 2, \qquad k_3 B = 1, \qquad k_2 = \tfrac{1}{10}.$$

This gives a steady state of $X_s = 10$, $Y_s = 20$, a small amplitude period of $T = 2\pi / \sqrt{2}$ and a constant time step of $\Delta t = \tfrac{1}{4}$ was again used.

The big variation in phase velocity on the outer orbit is most evident, and makes the time steps for too large for accurate results when X and Y are large. We have used the constant time increment $\Delta t = \tfrac{1}{4}$ simply to show how interesting qualitative results can be obtained in a matter of minutes on a pocket electronic calculator and also to model our evolution game in which the relatively large value of $\tfrac{1}{4}$ is necessary to make the game fast and interesting.

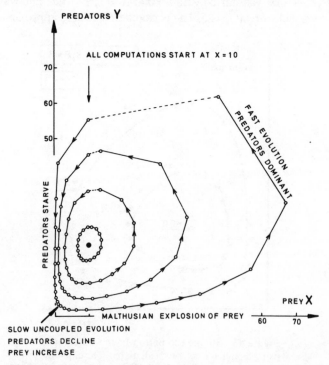

Figure 84 Crude numerical calculations modelling the proposed fox–rabbit evolution game

The trends of the outer orbits can be readily illustrated by an asymptotic analysis of the underlying Lotka–Volterra equations (1) as follows.

If X and Y are large we can ignore the linear terms on the right-hand sides to give

$$\dot{X} = -k_2 XY,$$

$$\dot{Y} = +k_2 XY,$$

so that

$$\dot{X} = -\dot{Y}.$$

The phase velocity is high and the curves regress at $45°$.

If, on the other hand, X and Y are small we can ignore the non-linear terms on the right-hand sides to give

$$\dot{X} = k_1 AX,$$

$$\dot{Y} = -k_3 BY.$$

The equations are now uncoupled and the rabbit population rises exponentially while the fox population decreases exponentially.

The underlying physical mechanisms are also apparent. Starting with low numbers of each species close to the origin $X = Y = 0$ the low number of foxes allows a population explosion of the rabbits following a roughly exponential Malthusian growth. Next, as their food becomes abundant we have an explosion in the fox population and they rapidly decimate the rabbit population. Once the prey reach a low population density, however, the foxes lack food and their numbers now decline, bringing us back to the starting point near the origin. This cycle can now be repeated indefinitely.

6.4 An evolution game

We now introduce an evolution game that brings an interesting and realistic random stochastic element into the finite difference scheme.

The game is designed around the finite difference equations (6) with as before,

$$k_1 A = 2, \quad k_3 B = 1, \quad k_2 = \tfrac{1}{10}, \quad \Delta t = \tfrac{1}{4}, \quad X_s = 10, \quad Y_s = 20,$$

so that

$$\Delta X = \tfrac{1}{2} X - \tfrac{1}{40} XY,$$

$$\Delta Y = \tfrac{1}{40} XY - \tfrac{1}{4} Y.$$

The rabbits, X, are represented by X white counters and the foxes, Y, by Y black counters.

The counters are thrown at random on to the board, adapted from a chessboard, shown in Figure 85. One-half of the total area of the board is white, one-quarter is grey, and one-quarter is black. The board is also more coarsely divided by the thick lines into 16 larger squares.

Figure 85 The modified chessboard on which the game is played

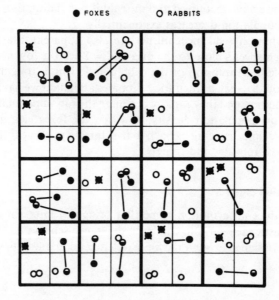

Figure 86 The results of a typical throw, showing dead foxes, breeding rabbits, and fox dinners

Any fox falling on a black square is *first* deemed to have died and is removed from the board. Since black represents one-quarter of the board this correctly models the mortality term $-\frac{1}{4}Y$, with, of course, a new random stochastic element.

Any rabbit falling on a white square is *secondly* deemed to have reproduced, one new rabbit being put on the board alongside it. Since white represents one-half of the board this correctly models the breeding term $+\frac{1}{2}X$.

Photograph 9 Helen studies the fate of the rabbits during the proposed evolution game played on a modified chessboard. Black counters represent the foxes and white counters represent the rabbits. After they are thrown at random onto the board close fox–rabbit encounters are regarded as fox dinners. A live predator looks on the game with disdain

Foxes are *finally* allowed to eat rabbits within one of the larger thick-lined squares. As many fox–rabbit pairs as possible are identified, taking each large square in turn. The rabbit of each pair is then replaced by a fox. This decreases the rabbit population and increases the fox population by equal amounts; and since the pairing is clearly dependent on the product of the current fox and rabbit populations we have correctly modelled the interaction terms provided the constant is correct.

In fact it was found in an experimental calibration that throwing repeatedly 20 foxes and 20 rabbits produced an average of about 10 interactions per throw giving

$$\Delta Y = 10 \qquad \text{for } X = 20, \ Y = 20$$

and hence a constant of $\frac{1}{40}$—which is why we chose this value in the first place!

The resulting new numbers of X and Y are *now* counted and re-thrown on to the board and the procedures are repeated. An example of the fox–rabbit pairing within the large thick-lined squares is shown in Figure 86.

One practical feature worth noting is that the board can be conveniently ruled on the inside of a box lid: the rim then prevents counters flying off the board (Photo 9). This is important as they must be dropped quite violently on to the centre of the board so as to distribute themselves fairly uniformly. If the counters are allowed to bunch at the centre of the board the probability of fox–rabbit encounters will be quite appreciably raised.

One or two game results are shown in Figure 87, and it is hoped that the interested reader might try the game for himself. Quite varied outcomes can be achieved, and one key decision left to the player is the choice of starting point.

The new realistic feature that emerges in the game is the possible extinction of a species—usually the rabbits! This arises when the population is low and a random 'fluctuation' is unfavourable, as shown in one of the plotted games.

Quite an interesting starting point is close to, or just above, the steady state (10, 20). Here the game starts slowly with small, erratic steps until a large fluctuation generates a more positive phase motion. From just above the steady state the rabbits usually decline, and interest centres on their possible extinction: if they survive while the fox population is declining they can yield a spectacular population explosion which often eventually exhausts the number of counters and the determination of the player. Sometimes a complete cycle can be achieved with a second touch-and-go extinction tussle for the rabbits.

Unfortunately, the foxes rarely become extinct, and players could experiment with the rules so that they also have an extinction struggle on each cycle. I shall be interested to hear about any such improvements!

An extra evolutionary element is to introduce as a mutant a baby super-rabbit with longer legs than its fellows. If paired with a fox this super-rabbit has a 50:50 chance of escape, to be decided by the toss of a coin. Super-rabbits breed super-rabbits when they fall on white squares, and interest centres on the success or failure of this favourable mutation.

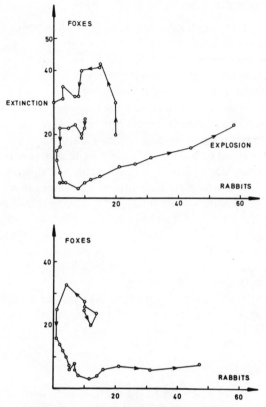

Figure 87 Some typical game results showing the possibilities of rabbit extinction or explosion

6.5 Structural stability

One acknowledged deficiency of the Lotka–Volterra equations noted by Nicolis and Prigogine[100] and May,[261] is that they have the structural instability of an *undamped* conservative mechnical system, the phase trajectories of which can be topologically changed by the introduction of infinitesimal viscous damping.

A more realistic set of equations could be expected to yield, not an infinity of neutrally stable trajectories but structurally stable *focuses* and *limit cycles* as with the chemical kinetics of the Brusselator. An attracting limit cycle would, for example, give rise to more coherent cyclic behaviour with a well-defined periodic time T.

Despite this lack of structural stability the usefulness of the Lotka–Volterra equations in predicting prey–predator oscillations in a remarkably simple manner is widely acknowledged.

6.6 Ecological considerations

Oscillating ecological systems have, of course, been observed in nature and Table 4 is taken from May's article.[261] Here K is an estimate of the environmental carrying capacity for the prey, and r/s is the prey/predator intrinsic growth rates. We notice that the hare–lynx system is the only one combining a large K with an r/s that is not small, which is May's *theoretical* condition for the existence of oscillations; and indeed, we see that it is the only system shown that is *observed* to exhibit cyclic behaviour. Table 4 is taken from an impressionistic summary of life history for eight natural prey–predator systems due to Tanner,[262] as reported by May. Perhaps we should have called it the hare and lynx game!

An accurate knowledge of the existing population dynamics is, of course, essential for good ecological planning and control. If, for example, in our game we wished to help the rabbits we might suggest removing a few foxes. If this were done from state A of Figure 88 it would yield a smaller cycle with approximately the same average number of rabbits. If it were thought that a large-scale destruction of foxes might help and the fox population were reduced to B, the result would be a rabbit explosion followed by a fox explosion followed by the almost certain *extinction* of the rabbits!

Such ecological studies take on a very real significance in the control of pests, such as locusts, and touch us even closer if the prey is man and the predator is smallpox.

6.7 Current work

We have already drawn attention to the authoritative work of May[106,261,263] and must finally mention his study of bifurcations and dynamic complexity of ecological systems.

Applications of catastrophe theory to evolution have been sketched by Thom[9], Waddington,[104] and Dodson;[105] and Zeeman[10] and Poston and Stewart[11] have

Table 4 A summary of the life histories of eight natural prey–predator systems due to Tanner,[262] as reported by May.[261] Reproduced with the permission of The Ecological Society of America (Copyright, 1975)

Prey	Predator	Geographical location	K large?	Ratio r/s	Behaviour
Sparrow	Hawk	Europe	No	2	Steady
Muskrat	Mink	North America	No	3	Steady
Hare	*Lynx*	*North America*	*Yes*	*1*	*Cycles*
Mule deer	Mountain lion	Rocky Mountains	Yes	0.5	Steady
White-tailed deer	Wolf	Ontario	Yes	0.6	Steady
Moose	Wolf	Isle Royale	Yes	0.4	Steady
Caribou	Wolf	Alaska	Yes	0.4	Steady
White sheep	Wolf	Alaska	Yes	0.2	Steady

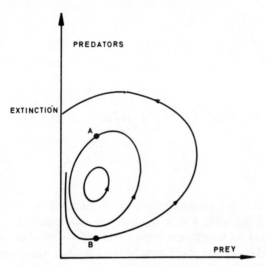

Figure 88   Possible mismanagement of an ecology by an apparently plausible strategy

both examined travelling waves and the stabilization of spatial frontiers, between, for example, grassland and forest. Poston and Stewart also discuss the 'economics' of bee colonies, including the advantages of aggregation.

Nicolis and Prigogine in their brilliant work on self-organization[100] study the thermodynamics and stability of evolutionary processes, including an analysis of pre-biotic polymer formation and evolutionary feedback. Organization in insect societies and the division of labour are also discussed.

CHAPTER 7

Hydrodynamic Instabilities and the Onset of Turbulence

We shall consider in this chapter bifurcational instabilities in the flow of viscous essentially incompressible fluids, looking in particular at the well-known Poiseuille flow down a long tube and the Couette flow between rotating cylinders. These are typical examples of the instability of fluid motions, very analogous to the buckling instability of engineering structures, in which a gradual increase of the Reynolds number leads to a loss of stability of the originally unique and stable primary flow. For Couette flow this first loss of stability can perhaps be seen as the first of a cascade of bifurcations that lead to the development of a fully turbulent flow at high velocities.

Following the discovery of a *strange attractor* by Lorenz,[145] it was suggested by Ruelle and Takens[146] that such attractors may play a key role in the development of hydrodynamic turbulence, and much modern research is directed towards this question. For this reason we present in the following chapter a simple example of a strange attractor due to Hénon.[158]

In the two cases of Poiseuille and Couette flow a particularly simple theoretical solution can be envisaged at low Reynolds numbers, which becomes unstable at a symmetric point of bifurcation. In the former flow down an *annular* pipe the mode of instability is *time periodic* and non-linearly *unstable*, while in the Couette motion between coaxial cylinders it is *steady* and non-linearly *stable*. For Couette flow we shall see how in a realistic experimental situation end effects destroy the symmetric bifurcation to give a tilted cusp that controls the structurally stable morphogenesis of cellular flows.

A superb general review of hydrodynamic turbulence, which lists many of the classical references including the pioneering work of Reynolds,[264] is given by Sir James Lighthill in a paper presented at the Osborne Reynolds Centenary Symposium in Manchester.[265]

7.1   Flow in a circular pipe

When a fluid such as water, which can normally be regarded as *viscous* but *incompressible*, flows down a pipe or round a curved obstacle, two quite distinct types of flow can be observed. If the fluid velocity is sufficiently low a steady *laminar* flow is observed, the fluid flowing smoothly down fixed streamlines. If,

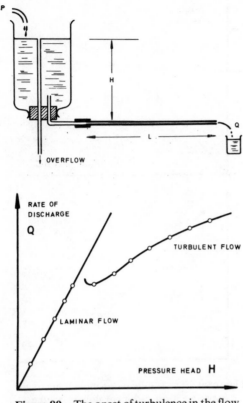

Figure 89   The onset of turbulence in the flow
of water down a pipe

however, the velocity is sufficiently high, an unsteady *turbulent* flow is developed in which local instantaneous velocities and pressures vary with time in an essentially random fashion.

Water running very slowly out of a tap is seen to be laminar, before surface tension breaks the accelerating and therefore thinning column, while a fast flow of water from the tap is readily seen to be turbulent.

A simple experimental study of this is easily made if water is driven down a thin circular pipe of length L by a pressure head H, as illustrated in Figure 89. If we measure the flow rate Q by, for example, noting the volume of water delivered in a suitable time interval, for various constant values of the head H we find that the graph of H against Q is originally linear but exhibits a discontinuous change as H and Q are increased beyond a critical value. This discontinuity in the response graph corresponds to the onset of turbulence, as can be assessed visually, and the precise critical values will be found to depend markedly on the care taken to avoid vibrational disturbances and to ensure smooth entry conditions, etc.

The results obtained from a simple home-made experiment performed with the help of my two children, Helen and Richard, are summarized in Figure 90 and Table 5.

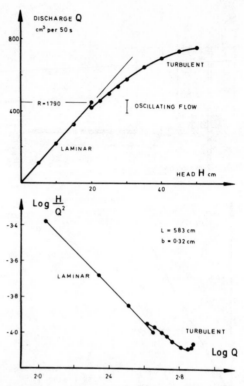

Figure 90 The results of a simple home-made
experiment on turbulent flow in a tube

Table 5 The results of a home experiment on the onset of turbulence in a plastic tube

Head H, cm	Q, cm³ per 50s	Log Q	Log H/Q^2
5	110	2.041	−3.383
10	220	2.342	−3.684
15	325	2.511	−3.847
20	450	2.653	−4.005
20	420	2.623	−3.945
22.5	460	2.662	−3.973
25	500	2.698	−4.000
27.5	540	2.732	−4.025
30	580	2.763	−4.049
35	650	2.812	−4.081
40	700	2.845	−4.088
45	740	2.869	−4.085
50	760	2.880	−4.062

Length of pipe = 583 cm
Radius of pipe = 0.32 cm
Kinematic viscosity of water was taken as 0.01 cm²/s.
The temperature was not recorded as it should always be in *laboratory* work because of the sharp
   variation of viscosity with temperature.

The experiment was performed by siphoning water from a bucket which was kept overflowing with water by a running hosepipe. The tube used was a plastic pipe of length 583 cm and internal radius approximately 0.32 cm, which was stretched tightly from its ends so that it hung in a smooth shallow curve. The quantity of water flowing in a 50-second time interval was measured using a graduated measuring flask for different values of the pressure head from the bucket surface to the horizontal outlet.

Away from the well-defined instability at a head of 20 cm the results could be repeated with great precision, but near to this value variable results were obtained and the flow was observed to pulse at a frequency of approximately 4 cycles per second, perhaps indicating that the flow was fluctuating between laminar and turbulent conditions. The high flow reading at $H = 20$ was the first measurement made at this head, and could not be repeated: no pulsations were observed, and it would seem that this was a lucky high-velocity laminar flow. Subsequently, at this pressure head the lower flow reading could be repeated many times.

The readings obtained are plotted directly in the top diagram of Figure 90 and a conventional Log–log plot is shown below. The Reynolds number at transition was calculated as 1790, but this must be regarded as a rather approximate figure because the internal diameter of the tube was not accurately assessed.

The flashes of turbulence observed in this home-made experiment are obviously similar to those discussed by Reynolds[264] and Lighthill.[265]

In the experiment of Figure 89 we can note that the pressure head H is the *controlled* quantity under which the fluid is free to adopt any rate of flow it wishes. This situation is, however, easily reversed, as in the dead and rigid loading of engineering structures in the laboratory, by enforcing a controlled flow rate Q with a piston or ram and then measuring the fluid's chosen pressure drop with a water manometer.

The Q–H diagram of Figure 89 is, moreover, highly reminiscent of the load-deflection diagram of a buckling cylindrical or spherical shell, a similarity which we shall see corresponds to a deep equivalence of the underlying instability phenomena.

We have seen that turbulence brings with it an *increased resistance* to flow, the magnitude of the flow rate Q for a given pressure head H being less than it would have been if laminar flow could have been preserved. So if we are concerned to drive water down a pipe, turbulence can be viewed as an undesirable property of the fluid motion, as it is in other practical situations.

The well-known theoretical Hagen–Poiseuille solution for the laminar flow of a viscous fluid down a circular pipe of radius b is readily derived. The fluid velocity at a radius r less than b, is written as $U(r)$ and the axial shear stress on a fluid cylinder of radius r is

$$\tau = \mu \frac{\mathrm{d}U}{\mathrm{d}r}$$

where μ is the coefficient of viscosity. If the pressure gradient driving the fluid is P, we can resolve axially for a fluid cylinder of unit length to obtain

$$\pi r^2 P + 2\pi r \mu \frac{dU}{dr} = 0$$

and hence

$$\frac{dU}{dr} = -\frac{P}{2\mu}r.$$

Integrating, we have

$$U = -\frac{P}{4\mu}r^2 + K$$

and since fluid in contact with the pipe wall is always at rest, the boundary condition

$$U(b) = 0$$

gives the constant of integration, K, as

$$K = \frac{P}{4\mu}b^2.$$

The velocity distribution over a cross-section of the pipe is thus given by

$$U(r) = \frac{P}{4\mu}(b^2 - r^2)$$

$$= U_m\left(1 - \frac{r^2}{b^2}\right)$$

indicating a parabolic variation of U with r, the maximum velocity at the centre-line being given by

$$U_m = U(0) = \frac{Pb^2}{4\mu}.$$

The mean velocity is readily found by integration to be exactly one-half of this maximum value, so the rate of discharge by volume is

$$Q = \pi b^2 \frac{Pb^2}{8\mu}$$

$$= \frac{\pi P b^4}{8\mu}.$$

This corresponds to the experimentally determined linear relationship between Q and H which exists under laminar flow conditions.

Now turbulence introduces fluid accelerations into the flow, so inertial forces will come into play and the equations of perturbed motion will include the density of the fluid, ρ, which we have not so far needed. These equations, and the simpler method of dimensions, suggest that with different fluids and similar experiments differing only in scale, turbulence will be initiated at a critical value of the dimensionless Reynolds number defined in general terms as

$$R = \frac{\text{characteristic velocity} \times \text{characteristic length}}{\text{kinematic viscosity}}$$

where the kinematic viscosity v is simply the ratio μ/ρ.

This conclusion is confirmed by experiments, and defining R specifically for our situation as

$$R = \frac{U_m b}{v}$$

the critical value is found to be of the order of 2200, as shown in Joseph's diagram (Ref. 266, page 121). In terms of the volumetric discharge this gives

$$R = \frac{2Q}{\pi v b} \doteq 2200.$$

We see that in a given apparatus a viscous liquid such as heavy motor oil will have a higher critical velocity than water. So at a given velocity water is more likely to be turbulent, as we would expect.

To examine this onset of turbulence in the Hagen–Poiseuille flow down a pipe it will now be convenient to consider the more general Poiseuille flow in an annular tube, which includes the present problem as a special case.

7.2   Flow in an annular tube

We consider, then, a viscous fluid flowing down the annular gap between an inner solid cylinder of radius a and a hollow coaxial outer cylinder of relevant radius b. This problem reduces to the earlier one if we simply set the radius a equal to zero.

Now it can be shown by a linear stability analysis based on the governing *Navier–Stokes* equations of fluid motion that the laminar flow down such an annular pipe loses its stability against infinitesimal disturbances at a critical value of the Reynolds number defined now as

$$R = \frac{U_m(b - a)}{v}.$$

This loss of stability corresponds to an *unstable-symmetric* point of bifurcation, which in fluid mechanics is usually termed a *sub-critical* branching point.

The mode of instability is *time periodic* as in the mathematicians' Hopf bifurcation. Such a time-dependent instability cannot of course arise in a conservative structural system or indeed in (elementary) catastrophe theory, but the phenomenon is familiar to engineers in the flutter of aerodynamically loaded wings and suspension bridges.

A non-linear analysis shows this incipient instability to be highly unstable, and on the response diagram of Figure 91 it gives the broken curve emerging from the

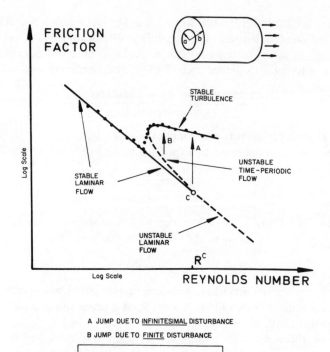

A JUMP DUE TO <u>INFINITESIMAL</u> DISTURBANCE

B JUMP DUE TO <u>FINITE</u> DISTURBANCE

$$R^C \rightarrow \infty \quad \text{AS} \quad \frac{a}{b} \rightarrow 0$$

Figure 91 The response diagram showing schematically the bifurcational triggering of turbulence in an annular pipe

critical branching point C, beyond which the laminar flow solution is unstable. This *schematic* diagram is based on the figures of Joseph (Ref. 266, pages 91 and 93), and the friction factor is a suitably chosen measure of the resistance to flow.

Thus if in an idealized experiment we could slowly increase the Reynolds number from a low value with only infinitesimal disturbances and a perfectly manufactured straight tube, the laminar flow solution would be maintained up to C. At this point, assuming that the Reynolds number is our controlled parameter, the flow pattern will jump suddenly to a state of stable but random turbulence, as indicated by the arrow A. This jump is *triggered* by the bifurcating time-periodic solution, but the end result is a turbulent motion seemingly unrelated to this incipient unstable motion. For this reason the complete *transition to turbulence* is thought to be either by *repeated branching*, in which there is a *successive loss of stability* of flows of less complicated structure to those with a more complicated structure,[266] or by a jump to a strange attractor.

Although stable against infinitesimal disturbances, the original laminar flow becomes increasingly precarious as the Reynolds number approaches R^C, and small-but-finite disturbances and imperfections can readily induce a premature jump to turbulence, as indicated by arrow B of Figure 91. This is confirmed by the experimental observation that very high Reynolds numbers can be observed in practice if great care is taken not to disturb the primary laminar flow.

The general scatter of most experimental transitions lies in the range $2000 < R < 2300$, which suggests that even under laboratory conditions there are large enough disturbances to knock the Reynolds number down to this plateau level. The results of one experimental study are indicated schematically by the solid circles in the figure.

We see that we have here a situation identical in many respects to the buckling of spherical and cylindrical shells in elastic stability. In each field we have a non-linearly *very unstable bifurcation* into a *triggering mode* that differs substantially from the dynamically adopted *final state*, and a severe *sensitivity* to imperfections and dynamic disturbances coupled with a fairly well-defined *plateau* of experimental failure levels. Indeed, the only qualitative difference between Figure 91 and our earlier Figure 25 of Chapter 1 for the buckling of a spherical shell is that in deference to the convention in fluid mechanics (and indeed the rest of science!) we have drawn the controlled parameter horizontally rather than vertically as is the perverse custom in shell buckling.

Our present story is completed by the observation that the critical Reynolds number of the linear theory goes to infinity as a/b goes to zero. So for flow in a circular tube with no inner core, arbitrarily high Reynolds numbers could be observed with sufficient care in the experimentation. The fact that most experimental transitions are still observed in the range $2000 < R < 2300$ attests

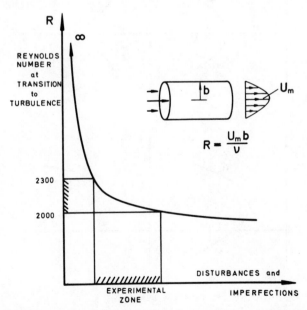

Figure 92   The infinite imperfection sensitivity of laminar flow in a pipe, showing the experimental plateau at $R = 2000$

to the fact that the Hagen–Poiseuille flow has a severe imperfection sensitivity coming down from infinity to a well-defined plateau, as illustrated schematically in Figure 92.

7.3 General hydrodynamic instabilities

As we have just seen for the particular example of Poiseuille flow down a tube, non-linear bifurcation problems in fluid mechanics have very precise analogies with those that arise in solid mechanics.

General theoretical analysis of the governing Navier–Stokes equations of motion for a well-defined flow, such as Poiseuille or Couette flow, shows that at low values of the Reynolds number there is a single solution corresponding to a *unique stable* steady laminar flow which we term the primary flow. This result relates closely, though not precisely due to its global nature, to the small-deflection Kirchhoff uniqueness theorem of linear elasticity.

Since this primary flow pattern cannot be observed experimentally at high Reynolds numbers, we must assume that it becomes unstable, at least against small-but-finite disturbances, with slowly increasing R. The study of this loss of stability is a central and fundamental problem in theoretical hydrodynamics that

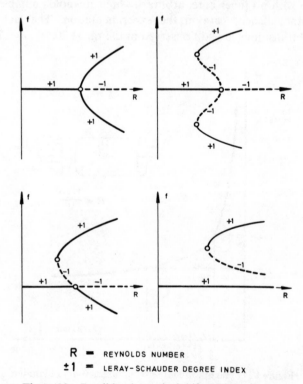

R = REYNOLDS NUMBER
±1 = LERAY–SCHAUDER DEGREE INDEX

Figure 93 Possible theoretical bifurcation diagrams for idealized steady fluid flows

has attracted a great deal of attention in recent years, paralleling the activity in elastic post-buckling.

The theoretical instability modes of fluids, predicted by bifurcation studies, are not usually those of fully developed turbulence with randomly fluctuating velocities and pressures, but are either *steady* or *time-periodic* secondary flows which can be thought of as the simplest possible turbulence representing a first step towards a fully turbulent regime. The existence here of time-periodic bifurcations emphasizes the lack of an energy potential in these problems.

Restricting attention now to steady flows, we shall present an outline of Benjamin's recent contribution.[50] A set of possible bifurcation diagrams for idealized theoretical flow situations is reproduced from Ref. 50 in Figure 93, the values of the Leray–Schauder degree index (± 1) being as shown. A negative value of this index guarantees instability of the corresponding flow solution, but, as with the stability determinant of conservative systems, a positive value does not guarantee stability. Benjamin has shown how the symmetric branching points of these diagrams, corresponding to the solutions of highly *idealized* theoretical models, are destroyed in real world experimental situations, and it is this feature that we shall now examine with particular reference to the Couette flow between rotating cylinders.

7.4 Taylor vortices between rotating cylinders

Consider then an essentially incompressible viscous liquid filling the cylindrical annular gap between two long solid coaxial cylinders as shown in Figure 94. If we fix the outer cylinder and rotate the inner cylinder with angular velocity ω the liquid will be given a circular motion. The liquid directly in contact with the inner cylinder will be dragged around with this cylinder, while the liquid in contact with the outer cylinder will always be completely at rest.

Under these conditions at low Reynolds numbers, and assuming for the time being that the cylinders are long in comparison to their radii so that end effects can be ignored, a particularly simple circular flow pattern will be generated. In this primary flow condition all particles of fluid will travel at constant speed around circular streamlines, the speed varying smoothly from ωr_1 at the inner boundary to zero at the outer boundary.

If we introduce cylindrical polar coordinates (r, θ, z) where z is measured axially, the corresponding velocities (v_r, v_θ, v_z) which in general might be written as functions of space and time as

$$v_r = v_r(r, \theta, z, t),$$
$$v_\theta = v_\theta(r, \theta, z, t),$$
$$v_z = v_z(r, \theta, z, t),$$

will simply be

$$v_r = 0,$$
$$v_\theta = v_\theta(r),$$
$$v_z = 0.$$

138

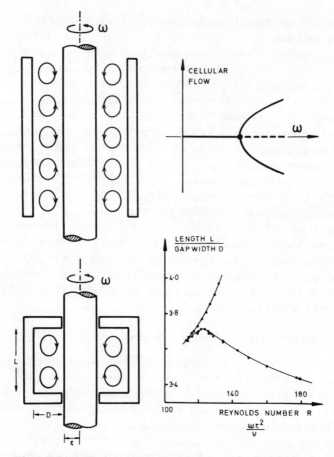

Figure 94 Taylor vortices between cylinders showing an idealized situation generating a stable-symmetric point of bifurcation and a practical situation giving a structurally stable tilted cusp. The experimental results of Benjamin[50] are reproduced with the permission of the Royal Society of London

Specifically the tangential velocity has the form (Ref. 267, page 8)

$$v_\theta = Ar + \frac{B}{r}$$

with the constants A and B given from the boundary conditions

$$v_\theta(r_1) = \omega r_1,$$
$$v_\theta(r_2) = 0.$$

As we increase the angular velocity, this primary flow may become unstable at a critical value of the Reynolds number which we define now as

$$R = \frac{r_1^2 \omega}{\nu}$$

and a *steady* secondary flow may be established in which velocities are periodic in the axial direction z. This is associated with cellular spiral vortex flows as illustrated in Figure 94, and the velocities can be written as

$$v_r = v_r(r, z),$$
$$v_\theta = v_\theta(r, z),$$
$$v_z = v_z(r, z).$$

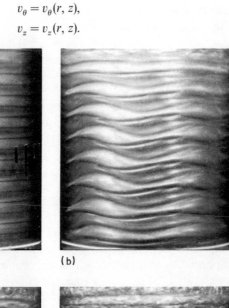

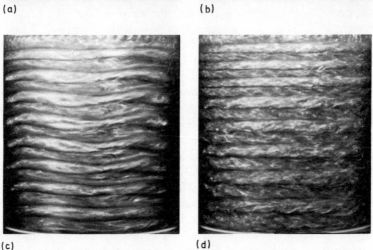

(a) (b)

(c) (d)

Photograph 10 The transition to turbulence in a fluid contained between concentric cylinders, with the inner cylinder rotating. (a) $R/R_c = 1.1$. Time-independent Taylor vortex flow with 18 vortices. The flows at the upper and lower fluid surfaces are inward. The vertical bars are fiducials separated by 10 degree angles. (b) $R/R_c = 6.0$. (c) $R/R_c = 16.0$. (d) $R/R_c = 23.5$. Figures (b) and (c) illustrate wavy vortex flow (with four waves around the annulus), while in (d) the waves have disappeared. In (b), (c), and (d) there are 17 vortices, and the flow is outward at the upper surface and inward at the lower surface. Reproduced with the permission of Cambridge University Press from P. R. Fenstermacher, H. L. Swinney and J. P. Gollub Dynamical instabilities and the transition to chaotic Taylor Vortex Flow, *J. Fluid Mech.*, **94**, 103 (1979)

We notice that the new secondary flow is *steady* in the sense that the velocities are not dependent on time t and rotationally symmetric in the sense that the velocities are not functions of θ. This steady secondary Couette flow was first observed by Taylor[268] in 1923, and beautiful photographs of the flow pattern due to P. R. Fenstermacher and H. L. Swinney are reproduced in Photograph 10.

A theoretical analysis of this idealized situation corresponding to an infinitely long cylindrical annulus shows that this loss of stability has the form of a *stable-symmetric* point of bifurcation (Figure 94) or in the terminology of fluid mechanics a *super-critical* bifurcation. For such a stable bifurcation we do not expect a severe sensitivity to dynamic disturbances or to the remote ends which strictly destroy the *simple* primary flow solution; indeed there is excellent agreement between experimental transition values of ω for long gaps with the critical value of the linearized stability analysis. Detailed comparisons between theoretical and experimental results can be found in Chapter VII of Chandrasekhar.[269]

This idealized situation is, however, essentially artificial, and for shorter annular gaps the effect of the ends will begin to dominate the flow morphology. For such conditions, Benjamin[50] invokes the concept of structural stability or genericity to argue that the symmetric bifurcation will be replaced by a tilted cusp observable if *two* control parameters are systematically varied, the cusp containing in its form an asymmetric (trans-critical) point of bifurcation.

To support this theoretical deduction, Benjamin presents the results of an experimental study of the morphogenesis of flow patterns, employing as his two controls the angular velocity of the inner cylinder and the length of the gap L. His apparatus was essentially the same as that employed by Taylor, and is shown schematically in the lower diagram of Figure 94. The length of the fluid-filled annulus was, however, comparatively small and could be continuously adjusted. The outer cylinder was made of perspex to allow observation of the fluid, an aqueous solution of glycerol rendered visible by the addition of a small quantity of a pearly substance. This outer cylinder was held fixed throughout and the inner cylinder was rotated at a controlled speed ω. The ratio of the cylidrical radii, r_1/r_2, was fixed at 0.615 and gap width, $D = r_2 - r_1$, was fixed at 23 mm. With this apparatus the end effects are strongly felt throughout the fluid, and steadily growing primary flows with two or four circulating cells could be observed.

The transitions between these two-cell and four-cell modes were studied in detail by the independent control of ω and L, and the cusp of Figure 94 was located corresponding to the flow transitions seen in Figure 95. This represents a stable cusp, the upper sheet corresponding to stable four-cell flows and the lower sheet to stable two-cell flows. The asymmetric trans-critical bifurcation can be observed under increasing Reynolds number at B.

The fine details of the cusp near C were reported to be difficult to resolve in the experiment (as is nearly always the case in experimental studies), but the quite definite hysteresis phenomenon for L/D values between B and C were repeatedly observed on increasing and then decreasing the angular velocity. This observed tilted cusp is completely analogous to that suggested by Zeeman[10] in de-

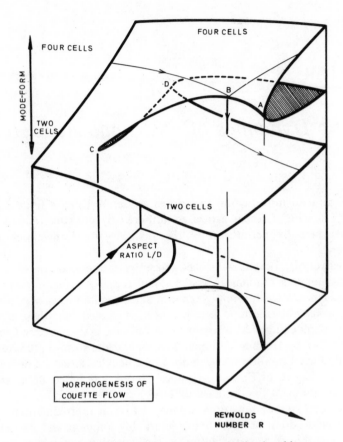

Figure 95 A schematic drawing of the inclined stable cusp
showing the morphogenesis of two-cell and four-cell forms

velopmental biology (Chapter 5), and represents a valuable step forward in our experimental observations of structurally stable morphogenesis.

Further recent studies of hydrodynamic turbulence are contained in the Proceedings of the Berkeley Turbulence Seminar,[270] and we note also the contribution of Kaplan and Yorke.[271] The use of catastrophe theory to elucidate the unfolding of an unstable persistently extensional flow including the creation of a vortex was demonstrated by Berry and Mackley.[84,85] The sudden changes of atmospheric circulation patterns, as modelled for example by a fluid annulus controlled by temperature contrast and rotation rate, have also been discussed in catastrophe theory terms by Lacher, McArthur, and Buzyna in *American Scientist* in 1977.

CHAPTER 8

Chaotic Dynamics of a Strange Attractor

We make now an introduction to strange attractors using a finite-difference mapping, following a recent article co-authored by my son Richard[272] and reproduced here by permission of the Institute of Mathematics and Its Applications.

The discovery by Lorenz of *strange attractors* in the response of extremely simple autonomous dynamical systems has recently aroused considerable interest since these attractors imply a chaotic and essentially *random* response of a well-defined *deterministic* model. It has been suggested that these attractors might offer a modelling of hydrodynamic turbulence, as discussed in Chapter 7, and recent work by Holmes has apparently identified a strange attractor in the sinusoidally forced non-linear oscillations of a buckled structure described by only a *single* degree of freedom. Holmes' study[151] is based upon analogue computer solutions of the ordinary differential equation and on analysis and digital computer studies of its Poincaré map P and an approximation, P_d, to P.

Such attractors do in fact seem to be remarkably common, and Rossler[147,148] has suggested that just as oscillation is *the* typical behaviour of dynamical systems with a two-dimensional phase space, so chaos might be the typical behaviour of systems with three or more phase dimensions. He defines chaos as characterized by 'an infinite number of unstable periodic trajectories and an uncountable number of non-periodic recurrent ones'.

Strange attractors may thus have a profound effect on our modelling of seemingly random behaviour, since it is now seen that a stochastic modelling may no longer be essential in all cases. For simple deterministic mechanical systems, they mean that computer results of their non-linear dynamics must be inspected with care (as must any results of a conventional Krylov–Bogoliubov averaging technique), since one feature of a strange attractor is that a sudden leap in response may occur after a long period of apparent quiescence. A strange attractor's known sensitivity to initial conditions must also be carefully watched, and new concepts of stability and repeatability are required.

In the present chapter we explore the iterative mapping of a plane onto itself, as studied by Hénon.[158] The mapping was chosen to simulate the Poincaré map of the Lorenz strange attractor,[145] and we explore different regions of Hénon's original picture using up to 10^6 iterations on a digital computer. This simple mapping can be explored more quickly and extensively than the three-dimensional continuous Lorenz system, but is shown to have essentially the same basic hierarchical properties as the latter. We extend Hénon's work by examining the divergence of the solutions from two closely spaced starting points, representing perhaps a simulated round-off error.

8.1 The finite-difference mapping of Hénon

Hénon has demonstrated, on the basis of numerical experiments, that a simple mapping of a plane onto itself *seems* to exhibit a strange attractor, similar to that discovered by Lorenz for a system of three first-order differential equations. This strange attractor appears to be the product of a one-dimensional manifold by a Cantor set.

For the Lorenz equations, the three-dimensional phase flow has a constant negative divergence, implying that any volume shrinks 'exponentially' with time. All trajectories tend to a set of measure zero, called an *attractor*. In some circumstances the attractor is simply a point, namely a stable equilibrium point, or a closed curve, namely a limit cycle. In other circumstances, however, the attractor has a more complex structure and is declared *strange*. Inside this strange attractor, trajectories wander in an erratic manner and are highly sensitive to initial conditions. This behaviour has stimulated much interest in dynamical systems theory and has been suggested as a possible model for hydrodynamic turbulence.

Hénon chose an iterative scheme to explore this type of behaviour quickly and accurately. Essentially he focuses attention, not on whole trajectories in the three-dimensional space but only on their successive intersections with a two-dimensional surface of section S. Thus as a trajectory leaves a given point A of S we follow it until it intersects S again at a new point $T(A)$, thus defining a mapping T of S into itself. Such a mapping is called a Poincaré map, and is illustrated in Figure 96. A trajectory is therefore replaced by an infinite set of points in S, obtained by the repeated application of the mapping T, and since the essential

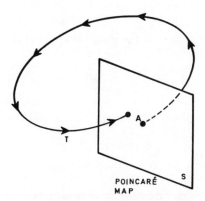

Figure 96 The Poincaré map of a trajectory in phase space

properties of the trajectory are preserved as corresponding properties of this set, the three-dimensional continuous problem is formally reduced to the study of a two-dimensional mapping.

Of course we cannot *find* the mapping of a given three-dimensional flow without the *integration* of the differential equations, but one approach is to guess an appropriate mapping and see how it behaves.

In this spirit, Hénon set out to determine a simple mapping that might realistically relate to a three-dimensional strange attractor. He thus suggested, on the basis of a scheme of folding, contracting, and rotating, the simple mapping T:

$$x_{i+1} = y_i + 1 - ax_i^2, \qquad y_{i+1} = bx_i$$

where a and b are constants. We note that the Jacobian of this mapping is

$$\frac{\partial(x_{i+1}, y_{i+1})}{\partial(x_i, y_i)} = -b$$

and so is a constant, the corresponding contraction of area for b less than one being the natural counterpart of the constant negative divergence of the Lorenz system.

The inverse of the mapping is readily written down, and we observe that T is a *one-to-one mapping* of the plane onto itself. This is a natural counterpart of the fact that there is a unique trajectory through any given point in the Lorenz system. This mapping, constructed empirically on the basis of its required properties, is shown by Hénon to be the most general quadratic mapping with constant Jacobian.

The mapping T has two invariant points given by setting $x_{i+1} = x_i$, $y_{i+1} = y_i$ as

$$x = \frac{-(1-b) \pm \sqrt{(1-b)^2 + 4a}}{2a},$$

$$y = bx.$$

These points are clearly real for

$$a > a_0 = \frac{(1-b)^2}{4}$$

and under this condition one of the points is always linearly *unstable*, while the other is unstable for

$$a > a_1 = \frac{3(1-b)^2}{4}.$$

The values chosen for the control parameters a and b were

$$a = 1.4, \qquad b = 0.3,$$

determined by Hénon after some experimentation.

8.2 Numerical results and magnification sequences

Independent of the (local) starting point, the iteration of T is found to give rise to the pictures of Figures 97 and 98, which presumeably therefore represent the

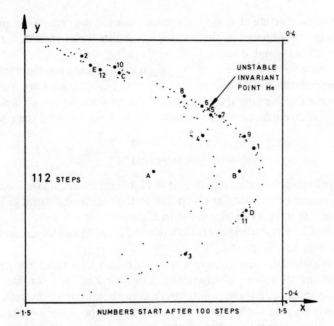

Figure 97 The strange attractor after 112 steps starting at the origin A

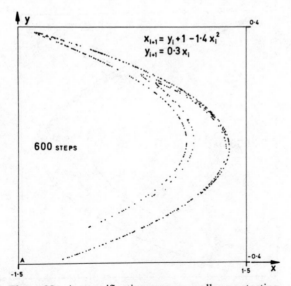

$$x_{i+1} = y_i + 1 - 1\cdot4\,x_i^2$$
$$y_{i+1} = 0\cdot3\,x_i$$

Figure 98 A magnification sequence, all runs starting at the point He. Using a not very economical programme these graphs took approximately 5, 18, 20, and 202 minutes respectively

146

attractor itself. Starting at $x_0 = 0$, $y_0 = 0$ for example, the iteration sequence A, B, C, D of Figure 97 shows that the point is lost within the attractor after only four iterations at this original scale.

Progressing with this run from the origin the attractor has been followed for 100 steps to damp out transients and the next twelve points are numbered to show the *random jumping* that typifies *all* the motions on the attractor.

Hénon observed that one of the two invariant points, with the coordinates

$$x = 0.63135448\ldots,$$
$$y = 0.18940634\ldots$$

appears to belong to the attractor, and to eliminate transients we have followed Hénon by normally starting at this point, to the indicated accuracy. This point, denoted by He, is marked as a cross in Figure 97.

A run of 600 steps starting at He is shown in Figure 98 and a similar run of 2000 steps is shown in Figure 99.

We next look in detail at the small box C of Figure 99 which is shown enlarged to the full size of Figure 100: here the computer has been run for 10^4 steps, plotting however only those points that fall in the square of interest. A similar magnification is made of box D which is enlarged to the full size of Figure 101 for 10^5 total steps. Here we see that what seemed to be a single curve in the earlier figures has been resolved into six parallel curves.

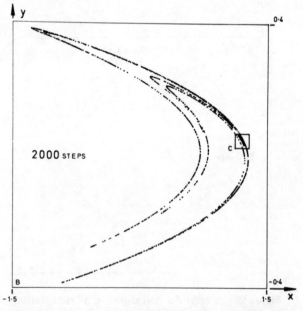

Figure 99 A magnification sequence, all runs starting at the point He. Using a not very economical programme these graphs took approximately 5, 18, 20, and 202 minutes respectively

147

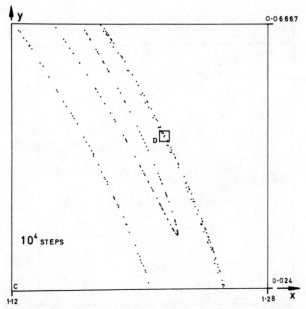

Figure 100 A magnification sequence, all runs starting at the point He. Using a not very economical programme these graphs took approximately 5, 18, 20, and 202 minutes respectively

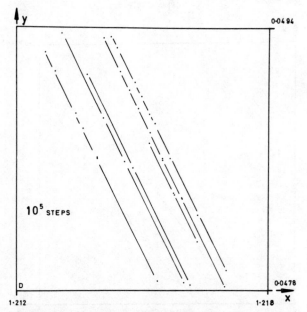

Figure 101 A magnification sequence, all runs starting at the point He. Using a not very economical programme these graphs took approximately 5, 18, 20, and 202 minutes respectively

148

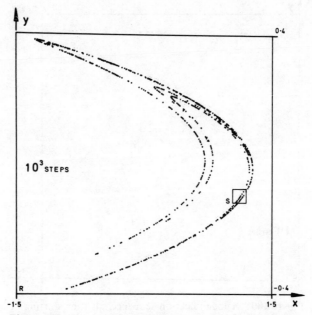

Figure 102 A second magnification sequence using a time – economy programme. The runs all start at He and took approximately 8 minutes, 9 minutes, and $12\frac{1}{2}$ hours respectively

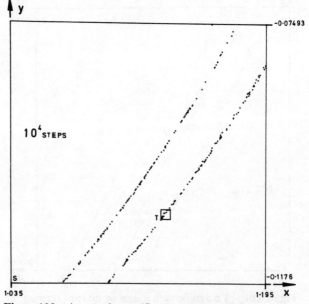

Figure 103 A second magnification sequence using a time – economy programme. The runs all start at He and took approximately 8 minutes, 9 minutes, and $12\frac{1}{2}$ hours respectively

A similar magnification sequence, starting again always at He, is shown in Figures 102 to 104. Here the final figure corresponds to 10^6 iterations, only a small fraction of which fall in the box and are plotted. Once again an original single curve is resolved into six parallel curves and we see clearly from the thickness of the six curves that a further magnification would result in an *even finer resolution.*

These 10^6 iterations took $12\frac{1}{2}$ hours on a Hewlett Packard 9815A desk-top computer linked to a 7225A graph plotter working to an accuracy of twelve digits.

Figure 105 shows the build-up of y with the number of iterations, starting nominally at the invariant point He. We see that after a slow start, the graph quickly develops a white noise character.

We see that the attractor in (x, y) space seems to consist of a number of roughly parallel 'curves', the points eventually distributing themselves densely over these curves in a quasi-random manner.

Transversely, there seems to be an indefinite multiplication of curves as we have seen, so that each apparent 'curve' is in fact made up of an infinity of quasi-parallel curves. On the basis of these observations, Hénon suggests that the transverse structure is that of a Cantor set.

The fact that after even our 10^6 steps the solution has not diverged suggests that there exists a region of the plane from which the points cannot escape, a

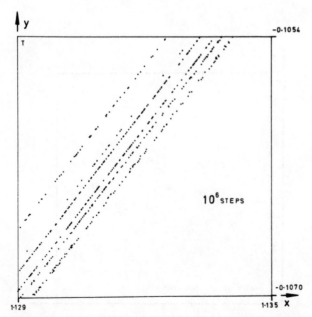

Figure 104 A second magnification sequence using a time – economy programme. The runs all start at He and took approximately 8 minutes, 9 minutes, and $12\frac{1}{2}$ hours respectively

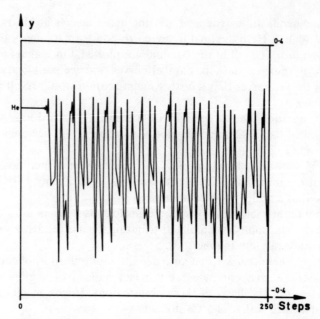

Figure 105 A plot of *y* against the number of steps, starting nominally at the invariant point He

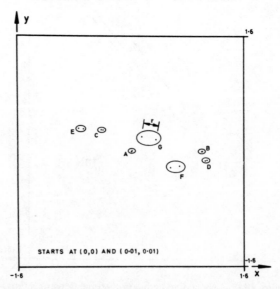

Figure 106 Initial divergence of two starts near the origin

result that is proved by Hénon by finding a region R around the attractor which is mapped inside itself.

8.3 Sensitivity to initial conditions

A feature of a strange attractor not specifically examined by Hénon is an extreme sensitivity to initial conditions, and we make here a preliminary study of this.

Figure 106 shows the divergence of two close starts during the first few iterations, and the corresponding plot of the separation r between the two solutions against the number of steps is shown in Figure 107. The points quickly move apart and generate a white noise appearance as their positions become essentially uncorrelated.

A similar graph for the very close starting points $(0, 0)$ and $(10^{-9}, 10^{-9})$ is shown in Figure 108 and the initial 'exponential' growth of the separation r is clearly seen.

To observe this phenomenon clearly a log plot is obviously required, and this is shown in Figure 109, where we see that for a close start we have roughly a straight line, as we might expect. Starts at $(0, 0)$ and $(10^{-K}, 10^{-K})$ are shown for values of K from 3 to 9, and we see that the graphs are well correlated until separations become relatively large and $-\log_{10} r$ tends to zero.

8.4 Sensitivity to computational round-off errors

The adjacent starts just discussed can be thought of as simulated round-off errors, and we see from Figure 109 that with a starting 'error' of order 10^{-3} we can only

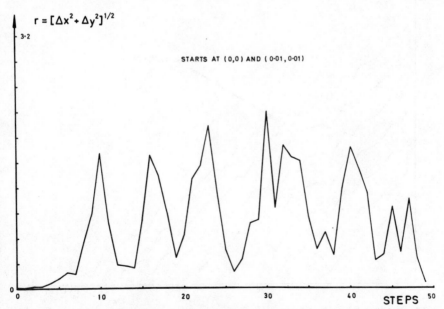

$$r = [\Delta x^2 + \Delta y^2]^{1/2}$$

3·2

STARTS AT (0,0) AND (0·01, 0·01)

STEPS

Figure 107 Plot of the separation r against the number of steps for two starts near the origin

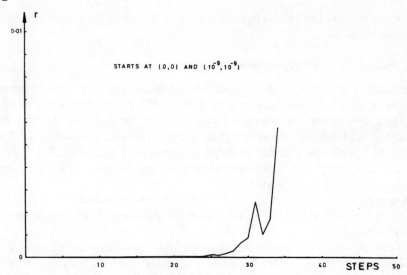

Figure 108 Plot of the separation r against the number of steps for two starts very close together near the origin

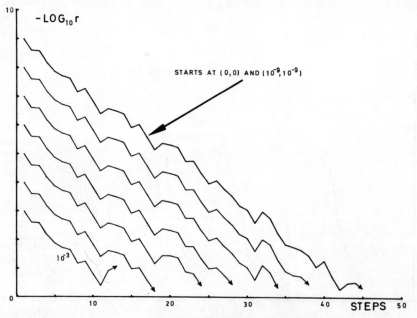

Figure 109 A log plot of the separation against the number of steps for a series of close starts near the origin

locate points accurately up to about 10 steps, while with a starting 'error' of order 10^{-9} we can locate points accurately up to about 40 steps. Clearly a computer of any size and accuracy is rapidly defeated by such extreme divergence: and Hénon's sixteen-digit accuracy on an IBM 7040 is really little better than our twelve-digit accuracy.

What meaning can we then attribute to the numerical solutions in the presence of this extreme divergence? The answer seems to be that while the correct position of the solution point *within the attracting curves* is rapidly lost, the location of the *curves themselves* is accurately determined by computations of this type. This is because, balancing the extreme divergence within the curves, there is an extreme convergence onto the curves of the attractor. This is why our curves look identical in overall shape to those computed by Hénon.

Whether or not Hénon has indeed generated a strange attractor has been questioned by some writers, but Holmes in a major recent article[151] suggests that Hénon is in fact right. That such an innocuous mapping can generate such complex and quasi-random behaviour is then of great interest to all computer users, a point that is only marginally weakened if the response proves to be merely a periodic motion of very long period.

8.5 Modern work and the chaotic motions of a buckled beam

Historically the strange behaviour of a simple deterministic dynamical system was first noted by Lorenz[145] in his work on atmospheric dynamics, while the suggested application to hydrodynamic turbulence was due to Ruelle and Takens.[146]

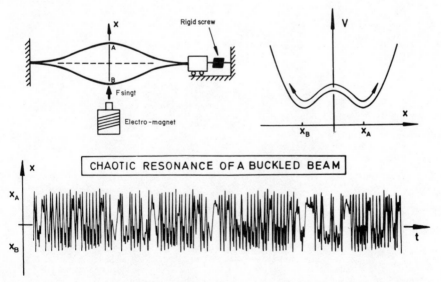

Figure 110 A trace of deflection against time for a sinusoidally excited buckled beam in a preliminary experiment

A number of significant modern papers on the subject are contained in the Annals of the New York Academy of Sciences edited by Gurel and Rossler.[135] A book on the subject by Gumowski and Mira[157] gives many pictures of the Hénon type, while several studies of the stability of finite-difference dynamical systems have been presented by Hsu and co-workers.[159-161]

An intensive study of strange chaotic behaviour in mechanics is due to Holmes.[149-156] He has examined the physical and computational implications of such essentially random behaviour and has made a specific study of a magnetoelastic system.

Particularly important is his detailed study of the chaotic motions of a slightly buckled beam subjected to a lateral sinusoidal excitation.[151] This beam can be modelled as a one-degree-of-freedom system with a highly non-linear potential energy defining the two close buckled configurations. The seemingly random vibrations of the beam involve a hesitant snapping action between the two available post-buckling equilibrium states, and a preliminary trace from an experimental study at University College is shown in Figure 110.

An excellent and very readable introduction to strange attractors, with pictures of the Hénon mapping, fluid turbulence, and the frequency spectra observed in Couette flow, is presented by Ruelle.[273] A strange attractor is suitably defined and a multitude of physical problems are discussed.

CHAPTER 9

Self-excited Oscillations Under Non-conservative Fluid Loading

A steady wind blowing around an engineering structure or water flowing rapidly through a pipe can pump energy continuously into the system, causing the build-up of large and dangerous vibrations and possibly collapse. Such fluid loading, which is not associated with a potential energy function like the $m\,g\,h$ of a mass m at height h in a gravitational field of strength g, is termed non-conservative. Here the lack of energy conservation that we have in mind is not the always-present dissipation of energy by internal damping but the availability of an energy *source*.

Once the loading on an elastic solid or structure is non-conservative in this sense, so that the applied forces are not derivable as the gradient of a total potential energy function, the possibility arises of dynamic bifurcations of the Hopf type and simple *static* and *energy* stability criteria breakdown.[35,37]

Much work has been done on problems of this type in civil, mechanical, and aero-space engineering, and a brief sketch of the field has been presented in Chapter 1 and in a corresponding publication.[274] Further useful references in the vibration and stability of mechanical systems can be mentioned here,[275-288] ranging from the classic work of Krylov and Bogoliubov in 1947 to the modern treatise of Nayfeh and Mook with its massive bibliography. Work on problems of dynamic buckling should also be noted.[289-291]

General non-conservative systems have been discussed by Plaut,[292,293] and an important step in the non-linear stability analysis of such systems has been made by Huseyin[294,295] using a perturbation approach. Huseyin employs a continuous transformation to a canonical Jacobian, akin to our earlier continuous diagonalization schemes,[36] to study the bifurcations to divergence and flutter of a general class of autonomous systems, the transformation allowing a complete stability discussion of all equilibrium paths involved. He also makes full use of multiple control parameters, following his extensive earlier work.[33,35] Dynamic bifurcations of the Hopf type have also been shown to be amenable to a treatment in the manner of catastrophe theory, as summarized briefly by Stewart in an important review.[296]

A general review of present and future instability problems facing the oil industry in the design of off-shore structures is given by Thornton to a recent engineering conference.[297] This covers the vortex-induced vibrations of moorings and riser pipes and the potential dynamic instabilities of floating platforms.

155

9.1 Aeroelastic galloping of a bluff structure

Three different forms of aeroelastic instability are unimodal *galloping, vortex resonance*, and bimodal *flutter*, and we shall be looking at the latter in the following section.

The spectacular oscillation and collapse of the Tacoma suspension bridge due to a steady wind,[298,299] shown in dramatic motion in Photograph 11, is now thought to be due to a combination of more than one of these phenomena. The H sections of some early suspension bridges were in fact particularly prone to vortex-induced torsional vibrations. The new Humber bridge of Photograph 12, currently the longest single span in the world, is more than one and a half times the length of the Tacoma suspension bridge, showing that the civil engineers have thoroughly mastered this early aeroelastic problem.

We look now at the mathematical treatment of pure galloping. Consider, for example, the wind of velocity V blowing past the square prism of Figure 26 of Chapter 1. The prism of mass m, height H, and length L is constrained to move only vertically at right angles to the wind, with a single generalized coordinate y

Photograph 11 The Tacoma Narrows bridge, Washington, exhibiting violent torsional oscillations in a steady wind. This suspension bridge with a span of 2800 feet collapsed in 1940 due to these aeroelastic vibrations only four months after its completion. It twisted about 45 degrees from the horzontal in both directions in a steady wind of about 42 mph before failure occurred. Photograph by courtesy of B. D. Elliot

157

Photograph 12 The new Humber suspension bridge has currently the longest single span in the world. With a span of 4626 feet it is more than one and a half times the length of the disastrous Tacoma bridge, showing that with careful attention to the aerodynamic design, the civil engineer has now overcome the earlier form of instability. The Consulting Engineers are Freeman Fox and Partners, and the photograph is reproduced with the permission of I. Innes

representing the *downwards* displacement from its equilibrium position. It is supported by an elastic spring of stiffness k and a viscous damper which supplies a retarding force $r\dot{y}$ always opposing the velocity $\dot{y}$.

Now when the prism moves downwards with velocity $\dot{y}$ the velocity of the air *relative* to the body is given by V_R in the triangle of velocities drawn. This relative wind velocity V_R at angle α will give rise to a vertical component of force

$$F_v = \tfrac{1}{2}\rho V^2 a C(\alpha)$$

where ρ is the air density, V is the prescribed *horizontal* wind velocity, and a is the frontal area HL.

Under a quasi-static assumption the coefficient C is simply dependent on the angle

$$\alpha = \tan^{-1}\frac{\dot{y}}{V},$$

and with the downwards force convention chosen we might indeed expect it to be *negative*, as it always is for *large* α. However, for *small* vertical velocities it can be positive for certain bluff profiles such as our square prism, and two typical variations of C with α are shown in the lower diagram. These graphs were determined by Parkinson and Brooks[95] from aerodynamic tests on *stationary* inclined sections following our quasi-static hypothesis. When C has the same sign as $\dot{y}$, the aerodynamic force *encourages* any initial motion, and the wind can be viewed as a negative damper.

The equation of motion of our aeroelastic oscillator is

$$m\ddot{y} + r\dot{y} + ky = \tfrac{1}{2}\rho V^2 a C(\alpha)$$

and from the experimentally determined $C(\alpha)$ diagram we can expand C as a power series in $\dot{y}/V = \tan\alpha$ to give the equation

$$m\ddot{y} + r\dot{y} + ky = \frac{1}{2}\rho V^2 a\left\{A_1\left(\frac{\dot{y}}{V}\right) - A_3\left(\frac{\dot{y}}{V}\right)^3 + A_5\left(\frac{\dot{y}}{V}\right)^5 - A_7\left(\frac{\dot{y}}{V}\right)^7\right\}.$$

Here we have written only odd powers due to the symmetry of our square cross-section. The positive and negative signs of the series are merely a matter of convention and ensure that for the square prism all the A_i coefficients are positive.

If we linearize this equation for very small oscillations we need retain only the first term of the C expansion, and the net effective damping multiplying $\dot{y}$ becomes

$$r - \frac{1}{2}\rho V^2 a\frac{A_1}{V}.$$

With A_1 *positive* for our square prism, this net damping becomes zero at the critical wind velocity given by

$$V^c = \frac{2r}{\rho a A_1}.$$

So for very small motions, any initial disturbance will be damped out in an oscillatory fashion for values of V less than V^C; but for V greater than V^C small disturbances will be magnified by the negative net damping and will give rise to exponentially growing oscillations. So the trivial equilibrium solution $y = 0$, valid for all V, becomes unstable at this critical wind velocity V^C.

Above this critical loading the linear theory predicts unlimited exponential growth of oscillations, but in reality the higher terms of the C expansion give rise to a limit cycle of finite amplitude. The size of this limit cycle grows from zero as V changes from V^C in the characteristic manner that we have seen for a dynamic point of bifurcation.

Some typical forms of the aerodynamic coefficient $C(\alpha)$ and the resulting dynamic instabilities are shown in Figure 111 due to Novak[97]. Here, since α is very closely equal to $\tan \alpha$ over the range of angles shown, the visual form of the $C(\alpha)$ graph is identical to that of a $C(\dot{y}/V)$ graph.

The first case representing a rectangular cross-section in a slightly turbulent wind has a $C(\alpha)$ curve whose slope decreases monotonically from an initial positive value. This gives rise to the *stable* dynamic bifurcation shown on a plot of

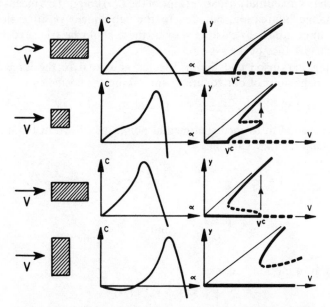

Figure 111 Four representative forms given by Novak[97] for the force coefficient $C(\alpha)$ together with the resulting dynamic response on a plot of the limit-cycle amplitude y against the wind velocity V. Dynamic bifurcations are shown at the critical velocities V^C, stable limit cycles are denoted by a continuous curve, while unstable limit cycles are represented by a broken curve. Jumps in vibration amplitude that would be observed during the slow increase of the wind velocity are indicated by the vertical arrows. Reproduced with the permission of the Science Council of Japan

the vibration amplitude y against the wind speed V. A stable limit cycle of amplitude y grows smoothly from zero as the wind speed increases beyond V^c.

The second situation is our previously discussed square prism in a steady wind. The slope of $C(\alpha)$ first decreases from its initial positive value giving rise once again to a *stable* dynamic bifurcation. After this initial decrease, however, the slope increases for a while before finally becoming negative. This temporary increase has a destabilizing action giving rise to the fold in the limit-cycle locus. Under slowly increasing V the system would thus experience a *jump* increase in the vibration amplitude as indicated by the jump arrow, and a hysteresis phenomenon would be observed during a subsequent decrease in wind speed. An experimental and theoretical study of this case will be made later in this section.

The third case of a rectangular block aligned in a steady wind has a $C(\alpha)$ graph whose slope initially increases giving an *unstable* dynamic bifurcation with a subsequent stabilization of the limit cycles as the slope finally decreases. Here again two dynamic jumps would be observed under the smooth variation of the wind speed V.

The final situation represents a rectangular block standing across a steady wind which has an initially negative slope of the $C(\alpha)$ graph. This means that there is no dynamic bifurcation, but due to the subsequent positive slope large-deflection limit cycles do exist as shown; these could be triggered by a *finite* dynamic disturbance of the system.

Following Parkinson and Smith[96] let us now analyse the large-deflection non-linear behaviour of our square prism, taking the coefficients

$$A_1 = 2.69, \qquad A_3 = 168, \qquad A_5 = 6270, \qquad \text{and} \quad A_7 = 59,900.$$

These give a good fit to the experimental points, as shown in Figure 112.

Introducing the new variables

$$Y = \frac{y}{H}, \qquad \tau = wt, \qquad w^2 = \frac{k}{m}, \qquad U = \frac{V}{wH},$$

$$n = \frac{\rho H^2 L}{2m}, \qquad \text{and} \qquad \beta = \frac{r}{2mw},$$

the equation of motion becomes

$$\ddot{Y} + 2\beta \dot{Y} + Y = nU^2 C$$

where a dot now denotes differentiation with respect to τ. This can be written more fully as

$$\ddot{Y} + Y = nA_1 \left\{ \left(U - \frac{2\beta}{nA_1} \right) \dot{Y} - \left(\frac{A_3}{A_1 U} \right) \dot{Y}^3 \right.$$

$$\left. + \left(\frac{A_5}{A_1 U^3} \right) \dot{Y}^5 - \left(\frac{A_7}{A_1 U^5} \right) \dot{Y}^7 \right\}$$

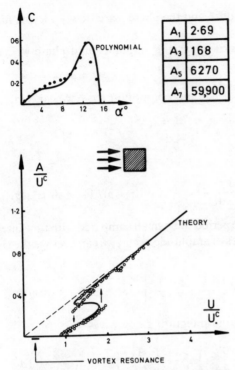

Figure 112 A polynomial fit to an experimentally determined $C(\alpha)$ characteristic for a square cross-section and the corresponding theoretical and experimental results of Parkinson and Smith[96] on a plot of vibration amplitude against wind velocity. The experimental points are seen to confirm the theoretically predicted hysteresis and final asymptote. The experimental wind velocities are seen to be well above the indicated region of vortex resonance. Reproduced with the permission of Oxford University Press

and we see that the critical value of the velocity parameter U is given by

$$U^C = \frac{2\beta}{nA_1}.$$

Our equilibrium equation has the form

$$\ddot{Y} + Y = F(\dot{Y})$$

and is solved using essentially the first approximation of Krylov and Bogoliubov by writing

$$Y = A\cos(\tau + p)$$

where the amplitude A and the phase p are slowly varying functions of the time parameter τ.

Subject to this slowly varying amplitude and phase we have approximately

$$\dot{Y} = -A \sin(\tau + p)$$

and multiplying our equation of motion by $\dot{Y}$ we can write

$$\frac{1}{2}\frac{d}{d\tau}(Y^2 + \dot{Y}^2) = \dot{Y}F(\dot{Y})$$

giving

$$\frac{1}{2}\frac{d}{d\tau}A^2 = -A \sin(\tau + p)F\{-A \sin(\tau + p)\}.$$

Assuming the cycle period to be small compared with time intervals during which appreciable changes of amplitude occur, we can average the right-hand side over a cycle and write

$$\frac{1}{2}\frac{dA^2}{d\tau} = -\frac{1}{2\pi}\int_0^{2\pi} A \sin vF(-A \sin v)\,dv.$$

Substituting for F and integrating now gives

$$\frac{dA^2}{d\tau} = nA_1\left\{\left(U - \frac{2\beta}{nA_1}\right)A^2 - \frac{3}{4}\left(\frac{A_3}{A_1 U}\right)A^4 + \frac{5}{8}\left(\frac{A_5}{A_1 U^3}\right)A^6 - \frac{35}{64}\left(\frac{A_7}{A_1 U^5}\right)A^8\right\}.$$

We can observe that we have essentially made a work balance, equating the increase of energy to the work done by the velocity-dependent forces. We note in passing that p, which has not appeared in this analysis, can be set equal to zero in the first approximation.

For a limit cycle of constant amplitude we can set the rate of increase of A^2 to zero to obtain

$$\left(U - \frac{2\beta}{nA_1}\right)A^2 - \frac{3}{4}\left(\frac{A_3}{A_1 U}\right)A^4 + \frac{5}{8}\left(\frac{A_5}{A_1 U^3}\right)A^6 - \frac{35}{64}\left(\frac{A_7}{A_1 U^5}\right)A^8 = 0.$$

Cancelling A^2 for the trivial solution and substituting for U^C we have finally

$$U - U^C = \left(\frac{3A_3}{4A_1}\right)\frac{A^2}{U} - \left(\frac{5A_5}{8A_1}\right)\frac{A^4}{U^3} + \left(\frac{35A_7}{64A_1}\right)\frac{A^6}{U^5},$$

indicating a dynamic bifurcation at $U = U^C$ with a first approximation to the post-critical behaviour as

$$U - U^C = \left(\frac{3A_3}{4A_1}\right)\frac{A^2}{U^C}.$$

Clearly the dynamic bifurcation is of the *stable* kind, with a growing stable limit cycle as U increases beyond U^C.

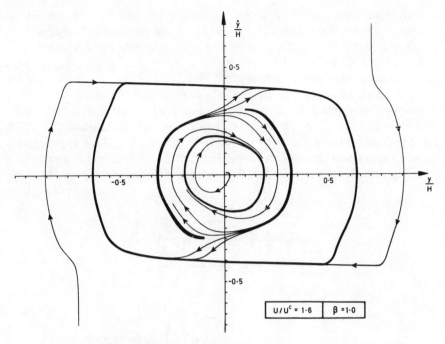

Figure 113 Two stable limit cycles separated by an unstable limit cycle

Dividing through by U we have

$$1 - \frac{U^C}{U} = \left(\frac{3A_3}{4A_1}\right)\left(\frac{A}{U}\right)^2 - \left(\frac{5A_5}{8A_1}\right)\left(\frac{A}{U}\right)^4 + \left(\frac{35A_7}{64A_1}\right)\left(\frac{A}{U}\right)^6$$

and we see that as U tends to infinity the left-hand side tends to unity, so A/U tends to a constant value. This means that a plot of A against U will tend to an asymptote through the origin. We also observe that we shall have a unique curve independent of n and β if we plot A/U^C against U/U^C.

The results of this calculation are shown in Figure 112, where they are seen to be in excellent agreement with the experimental results also reported by Parkinson and Smith.[96] The stability of the limit cycles can be assessed by studying the sign of $dA^2/d\tau$, and the experimental hysteresis is clearly seen.

A phase portrait showing a nest of three limit cycles at $U/U^C = 1.6$ is shown in Figure 113, and an alternative plot to obtain a universal curve is described by Novak,[300] who also considers the galloping behaviour of continuous systems. A useful review of the literature is contained in the recent book by Blevins.[98]

9.2 Aircraft panel flutter at high supersonic speeds

Flutter instabilities of plate and shell panels represent an important technical problem, stimulated by associated structural failures of high-performance

164

aircraft, spacecraft, and jet engines.[111] Modern research has done much to overcome original discrepancies between theory and experiment, but important non-linear problems remain. It is in fact predominantly the structural non-linearities of a plate that stabilize the incipient flutter at a finite limit cycle. This stabilization usually prevents immediate structural failure, as often arises for lifting surfaces, but opens the way to long-term fatigue failure.

Extensive experimental studies using supersonic wind tunnels have been made, as well as flight tests on the X-15 hypersonic research aircraft and the S-IVB launch vehicle for the Apollo moon exploration programme. These, together with Dowell's non-linear theories,[110] show that supersonic flutter corresponds to a stable dynamic Hopf bifurcation, as illustrated schematically in Figure 114. Here,

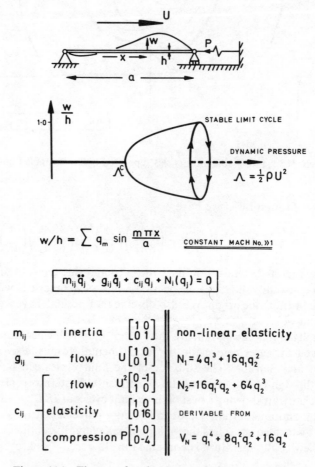

Figure 114 Flutter of a beam panel due to a high supersonic flow. The basic control parameter is the dynamic pressure Λ with the Mach number held constant at a value considerably greater than one. A second control parameter is the pre-stress P

as is conventional, the control parameter is taken as the dynamic pressure $\Lambda = \frac{1}{2}\rho U^2$, half the product of the density and the square of the flow velocity, while the Mach number, considerably greater than one, is held constant.

Specifically, let us consider[110] a high supersonic flow past the panel of Figure 114. Here we have a pin-ended beam of length a and an in-plane constraining spring which can be used to pre-stress the panel.

Dowell uses elasticity equations corresponding to the Von Karman large deflection plate equations, so the structural non-linearity is due to in-plane membrane action rather than curvature effects. Quasi-steady and linear aerodynamic theory is employed giving normal forces proportional to the slope $\partial w/\partial x$ and the rate $\partial w/\partial t$. We might note here that *non-linear* aerodynamic forces have been considered by Bolotin.[280]

A Fourier series is assumed for $w(x)$ of the form shown, and we shall here consider only the first two terms. Dowell considers six modes, which are necessary for accurate quantitative results, but he shows that two modes give a correct qualitative picture.

A Galerkin procedure is used to obtain the equations of motion in the form of the boxed equation, with the contributions listed below the box. These equations are then solved by numerical integration. Here the structural inertia terms are those of simple beam vibration. The analysis includes no structural damping, but the fluid flow gives us an effective positive definite viscous dissipation proportional to U as shown for g_{ij}. The fluid flow also gives displacement-dependent c_{ij} forces which are *antimetric* and therefore *circulatory*, varying as the square of the velocity.

We have next the well-known elasticity matrix for a beam and the usual destabilizing matrix associated with the pre-compression P. Finally, the non-linear structural elastic contributions N_1 and N_2 are derivable from the non-linear potential V_N; these can be derived from a large deflection beam formulation[36] using the square of the end-shortening.

Since the rate-dependent fluid forces correspond to positive definite viscous dissipation, we see that the *direct cause* of this classical flutter instability is the *circulatory matrix* c_{ij}. A comparison with other causes of flutter is shown in Figure 30 of Chapter 1. We observe further that the linear instability is satisfactorily stabilized by the non-linear *elastic* membrane action.

Dowell gives finally the interaction diagram for flutter and divergence in the control space of Λ and P, and describes some fascinating non-linear interactions between the two phenomena. This aspect has been followed up by Holmes[113] who relates it to the classifications of Takens.[301,302] Dowell also studies the effect of a static pressure differential which we have here ignored.

Flutter studies published by the Aeronautical Research Council due to Done should finally be mentioned here.[303,304]

9.3 Static and dynamic instabilities of a pipe conveying fluid

When a fast stream of water flows down a flexible elastic pipe or tube it exerts resultant lateral forces on the pipe along its whole length.

If the tube is stationary but curved, the fluid experiences centripetal accelerations of magnitude U^2/R, where U is the flow velocity and R is the local radius of curvature of the pipe. There is thus a lateral centrifugal force on an element of pipe of length δx of magnitude $MU^2\delta x/R$, where M is the mass per unit length of the fluid. Here M could have been written as ρA, where ρ is the density of the fluid and A is the cross-sectional area of the pipe.

These lateral centrifugal forces distributed along the length of the pipe are, surprisingly, mechanically equivalent to a compressive end load of magnitude MU^2 at a discharging tip.[305] It should be remembered, however, that *physically* there is no such end load in the absence of a nozzle.

For a simply supported pipe which can have no lateral deflection at either end, and one end of which is discharging the fluid into the atmosphere, this back reaction MU^2 is essentially *conservative*. It can buckle an initially straight elastic pipe at the Euler buckling load given by

$$MU^2 = \frac{\pi^2 EI}{L^2}$$

the right-hand side having been derived in Chapter 2 for mechanical loading. Here EI is the bending stiffness of the pipe and L is its length between pinned supports. This static buckling, called divergence to distinguish it from oscillatory dynamic flutter, is identical to that observed in the mechanically loaded column, and corresponds to a non-linear stable-symmetric point of bifurcation at which the trivial equilibrium path of no deflection intersects a post-buckling supercritical and stable equilibrium path.[305]

For a cantilevered pipe, clamped at one end and discharging into the atmosphere at the other free end, the back reaction MU^2 is always tangential to the tip. Such a *follower-force* is non-conservative and cannot be derived from a potential energy function. It is classified as circulatory, as we have sketched in Chapter 1, and it can pump energy into the pipe during any closed oscillation.

Considering an initially straight cantilevered elastic pipe, this end force cannot hold the pipe in a static deflected state, as becomes clear if we consider the sign of the curvature and the bending moment near to the tip. However we try to draw the pipe, the curvature and moment are in opposite senses, which is of course not consistent with the elastic properties of the pipe. So for a cantilevered pipe there are no equilibrium solutions at all, at either large or small deflections, apart from the straight trivial fundamental state.

A cantilevered pipe therefore exhibits no static bifurcation, but it is a common observation that the straight configuration of such a pipe becomes unstable at a certain flow velocity when the pipe develops dynamic oscillations of large amplitude. A length of flexible rubber tubing attached to a tap will exhibit this dynamic *flutter*. This instability corresponds to a dynamic Hopf bifurcation at which a linear analysis predicts an exponentially increasing oscillation, and Benjamin's study of an articulated model pipe has been discussed in Chapter 1.

Once a pipe (simply supported or cantilevered) *moves*, a second distributed lateral force comes into play due to the *coriolis* acceleration of the fluid flowing through a rotating pipe element. This comes from the polar-coordinate acceleration $2\dot{r}\dot{\theta}$, where $\dot{r}$ is here the fluid velocity and $\dot{\theta}$ is the rate of rotation of the element; the role of this acceleration in centrifugal laboratory testing is explored in a short paper.[306]

For the small deflections of a simply supported pipe these coriolis forces are gyroscopic in nature and do no work in a *real* pipe movement; they will, however, do work in certain virtual movements and must be included in any dynamic analysis. For the small deflections of a cantilevered pipe these coriolis forces do negative work on the pipe, and so dissipate energy, as can be seen in Figures 28 to 30 of Chapter 1.

A third force from the flowing fluid to the pipe is the *axial* frictional drag, but this usually drops straight out of an analysis[305] and plays no role in the large or small deflection (statics or dynamics) of a pipe, as argued carefully by Benjamin.[114]

The differential field equation for the small deflection $w(x, t)$ of an elastic pipe is shown as the sum of the lateral forces on an element in the first column of Figure 115. Here w is the lateral deflection at time t of a point originally distance x from the left-hand support. This differential equation of motion applies to both simply supported and cantilevered pipes, since the distinction between these two cases arises only through the different boundary conditions.

The first term represents the normal *inertial* force arising from the mass per unit length of pipe and fluid, m and M respectively. These multiply the lateral acceleration $\partial^2 w/\partial t^2$.

The second term represents the viscous damping of the air surrounding the pipe, being the product of a damping coefficient c and the lateral velocity $\partial w/\partial t$.

The third term represents the damping due to the viscoelastic nature of the pipe material. Here E^*I is a pipe parameter, akin to the bending stiffness EI.

The fourth term represents the aforementioned coriolis force, the cross-derivative $\partial^2 w/\partial x\,\partial t$ being the rate of rotation of an element. We notice that this term is linear in the flow velocity U.

The fifth term is the normal bending term from elasticity theory, arising from the shearing force within the pipe. The derivation of this term is sketched below the view of an element.

The sixth term represents the destabilizing action of a supposed pre-compression P_0. It arises as a result of the membrane action of the compression on the curvature $\partial^2 w/\partial x^2$.

The seventh term arises if the pipe is supported on a continuous elastic foundation of stiffness K. Such a foundation can be supposed to give a lateral force per unit length of Kw.

The eighth term is the contribution from the centrifugal forces already discussed, being MU^2 times the linear approximation to the curvature, $\partial^2 w/\partial x^2$. We see here how this looks entirely equivalent to the action of the pre-compression P_0.

The ninth and tenth non-linear terms arising from an assumed axial constraint need not concern us here, and equating the sum of the first eight terms to zero

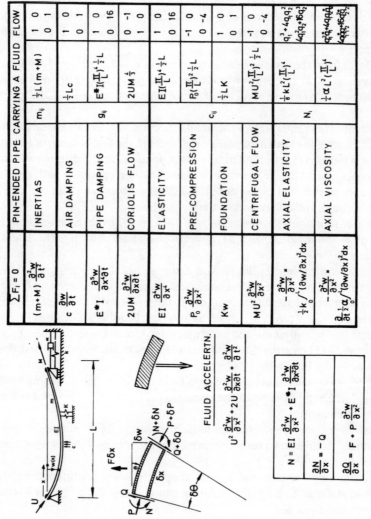

Figure 115. Diagram and coefficients for the dynamics of a simply supported pipe conveying a fluid. The sum of the lateral forces on an element gives the differential equation of motion, which is here discretized by a two-harmonic Galerkin procedure

gives us the linear differential equation of motion for a pipe undergoing small lateral vibrations. The derivation of the linear equation, excluding just the foundation, can be found in the paper of Paidoussis and Issid.[307]

Writing our differential equation of motion as

$$D\{w(x)\} = 0$$

we now proceed to solve it approximately for the *simply supported* pipe using the Galerkin procedure. To do this we approximate $w(x, t)$ by a two-term Fourier series

$$w(x, t) = q_1(t)\sin\frac{\pi x}{L} + q_2(t)\sin\frac{2\pi x}{L},$$

each term of which satisfies the simply supported boundary conditions for a pin-ended beam or pipe.

Substituting this expression into D the result is not zero, but we now set two weighted x averages of D equal to zero, choosing the modes themselves as the weighting functions, by writing

$$\int_0^L D\{w(x)\}\sin\frac{\pi x}{L} = 0,$$

$$\int_0^L D\{w(x)\}\sin\frac{2\pi x}{L} = 0.$$

Performing the integrations and making use of the well-known Fourier orthogonalities, we obtain after some algebra the pair of coupled ordinary differential equations

$$m_{ij}\ddot{q}_j + g_{ij}\dot{q}_j + c_{ij}q_j = 0$$

where the dummy-suffix summation convention is employed with all summations ranging from 1 to 2 and a dot denotes differentiation with respect to the time t. The contributions to the coefficients m_{ij}, g_{ij}, and c_{ij} are given in the final columns of Figure 115.

We notice that, for the energy terms, the results of the Galerkin procedure are exactly the same as we would have obtained using the Rayleigh–Ritz procedure. The air and pipe damping give us symmetric positive definite matrices, the former having the form of the mass matrix and the latter the form of the elasticity EI matrix. The well-known symmetric forms of the elasticity and pre-compression matrices give us the two lowest Euler loads for a strut, the pre-compression being augmented by the MU^2 of the centrifugal forces. The elastic foundation gives us a simple diagonal matrix, allowing the Euler loads of a strut on an elastic foundation to be easily derived, as we have seen in Chapter 2.

The coriolis forces are seen to give us an antimetric matrix. These forces are thus *workless* and can be classified as *gyroscopic*; they introduce the only linear coupling between the two differential equations of motion. Since we have positive definite damping in this gyroscopic system, a minimum of the total potential energy is both *necessary and sufficient* for stability, so the instabilities can be predicted in a simple manner. Limit cycles are impossible due to the continuous dissipation of energy, and hence a dynamic Hopf bifurcation cannot occur. However, as we shall see, increasing oscillatory motions can be generated by the coriolis forces about the trivial unstable equilibrium state above the second Euler critical load. In other words, although they can influence the detailed dynamics, the workless coriolis forces do not change the stability regions of the *damped* system.

Let us study first the linear problem *without* the gyroscopic forces. This will give us the *stability information* of the flow-loaded pipe, while if we set U and M equal to zero, we shall have the *real dynamics* of a damped Euler column, with or without an elastic foundation.

Crossing out the coriolis terms, the equations of motion are uncoupled, and we are left with the equations of two independent damped linear oscillators. The damping is always positive, and so we have just static buckling (divergence) when the stiffnesses vanish. So setting c_{11} equal to zero gives us buckling in mode one when

$$P_0 + MU^2 = EI\left(\frac{\pi}{L}\right)^2 + K\left(\frac{L}{\pi}\right)^2$$

and setting c_{22} equal to zero gives us buckling in mode two when

$$P_0 + MU^2 = 4EI\left(\frac{\pi}{L}\right)^2 + \tfrac{1}{4}K\left(\frac{L}{\pi}\right)^2.$$

Here the fluid flow is increasing the effective compression, by MU^2, and the effect of the elastic foundation can be observed. With $M = U = K = 0$ we retrieve the usual Euler buckling formula.

We consider secondly the linear behaviour of the flow-loaded gyroscopic system, ignoring for now the damping terms. This means that we are dealing with a *pathological* system in which any gyroscopic stabilization would not be preserved on the introduction of even infinitesimal damping, as we shall see later.

The linear equations of motion are

$$m_{ij}\ddot{q}_j + g_{ij}\dot{q}_j + c_{ij}q_j = 0$$

and setting

$$q_j = A_j e^{\lambda t}$$

the condition for a non-trivial solution is that the characteristic determinant must be zero,

$$|m_{ij}\lambda^2 + g_{ij}\lambda + c_{ij}| = 0$$

Now m_{ij} and c_{ij} are diagonal, and with no damping g_{ij} is antimetric, so the determinental equation becomes

$$(m_{11}\lambda^2 + c_{11})(m_{22}\lambda^2 + c_{22}) + g_{12}^2\lambda^2 = 0$$

and writing $\lambda^2 = N$ this becomes

$$N^2(m_{11}m_{22}) + N(m_{11}c_{22} + m_{22}c_{11} + g_{12}^2) + c_{11}c_{22} = 0.$$

To get the coefficients in a convenient dimensionless form we now divide through by the mass coefficient, and secondly introduce a fictitious time $\tau = zt$, choosing z to make the elasticity coefficient unity. The differential equations of motion, with a dot now denoting differentiation with respect to τ can then be written as

$$\begin{bmatrix} 1 & 0 \\ 0 & 1 \end{bmatrix} \ddot{q}_j + \sqrt{\alpha\Lambda} \begin{bmatrix} 0 & -1 \\ 1 & 0 \end{bmatrix} \dot{q}_j + \begin{bmatrix} 1 & 0 \\ 0 & 16 \end{bmatrix} q_j + \Lambda \begin{bmatrix} -1 & 0 \\ 0 & -4 \end{bmatrix} q_j = 0$$

where

$$\alpha = \left(\frac{16}{3\pi} \right)^2 \frac{M}{m + M}$$

and we have as a loading parameter

$$\Lambda = \frac{MU^2}{\pi^2 EI/L^2}.$$

Here we are focusing attention on the *fluid* loading of a pipe without an elastic foundation by setting $P_0 = K = 0$.

Taking the coefficients from this normalized equation our characteristic determinental equation becomes

$$N^2 + N(17 - 5\Lambda + \alpha\Lambda) + (1 - \Lambda)(16 - 4\Lambda) = 0.$$

We notice that this is satisfied by $N = 0$ when the loading corresponds to Euler buckling at Λ equal to one or four.

In this equation α represents the relative strength of the coriolis force, being zero when $M/m = 0$ and taking its maximum value of $(16/3\pi)^2 = 2.882$ when M/m tends to infinity.

To examine the stability of the trivial equilibrium state of the pipe, it is now necessary, for a fixed value of α, to study the movement of the roots of the characteristic equation as the loading parameter Λ is increased from zero. Writing

$$\lambda = R + Ii$$

where i is the square root of minus one, this movement is *plotted* in the three-dimensional space of (Λ, R, I) in Figure 116.

For $\alpha = 0$ we have the behaviour of an Euler column, shown in the first diagram: the two static instabilities here arise when the roots successively change from imaginary to real at $\Lambda = 1$ and 4.

For $\alpha = 0.03$ the only noticeable difference is the development of two closed loops where the relatively small coriolis force can induce oscillatory behaviour in a small region of roughly equal real parts.

172

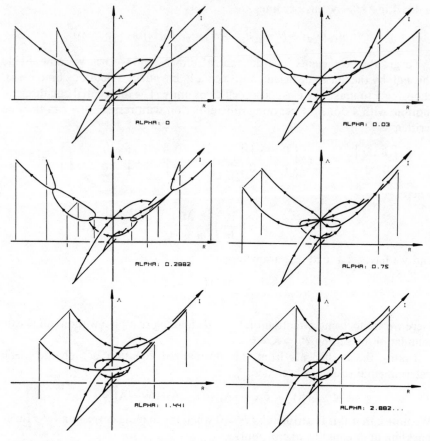

Figure 116 Movement of the roots in the complex Argand plane (R, I) shown as a three-dimensional picture with the Λ load axis vertical. These are drawn for a simply supported pipe with no damping

For $\alpha = 0.2882$ at $M/(M + m) = 0.1$ these closed loops have grown, and the curves are shown more clearly for low Λ in the *sketch* of figure 117 where some salient features have been somewhat emphasized. The onset of the flutter-like growing oscillatory motions at F can be clearly seen. These 'flutter' points F move inwards with increasing α, until at $\alpha = 0.75$ we have their coincidence with the $\Lambda = 4$ static bifurcation as shown in Figure 116.

For the higher value of $\alpha = 1.441$ corresponding to $M/(M + m) = 0.5$ the plot is shown in Figure 116 and an exaggerated sketch is shown in Figure 117. The second static bifurcation at $\Lambda = 4$ now corresponds, not to the $I \to R$ progression of the second roots but the reversed $R \to I$ progression of the first roots. This means that we have a region of *stability* above the second Euler load before the onset of 'flutter' at F. This *gyroscopic stabilization* should, however, be described as *temporary* since it is completely destroyed even by infinitesimal damping. This

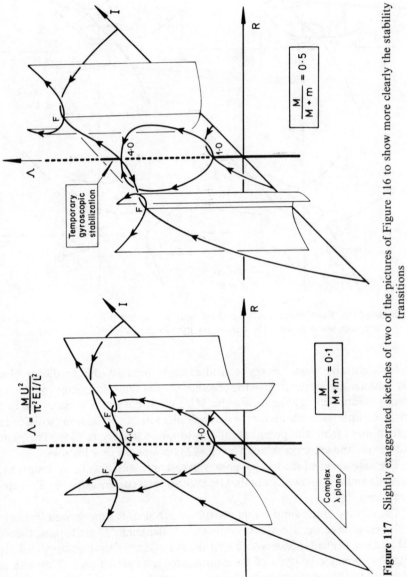

Figure 117 Slightly exaggerated sketches of two of the pictures of Figure 116 to show more clearly the stability transitions

174

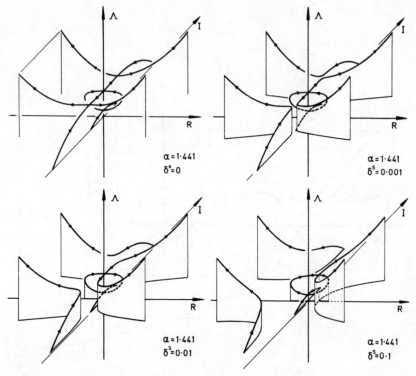

Figure 118 Root movements for the simply supported pipe in the presence of damping, showing how the temporary gyroscopic stabilization is destroyed

follows directly from energy considerations, because the predicted closed oscillations are not possible in the presence of even small damping; and since the total potential energy (including the MU^2 contribution) is a *maximum* in the trivial equilibrium state above the second Euler load, light damping will drive the system *away* from the potential hill-top in an oscillatory fashion. The plotted picture for the extreme value of $\alpha = 2.882$ contains no new features.

The manner in which the temporary gyroscopic stabilization is destroyed by small damping is shown in Figure 118, taken from a current paper by Thompson and Lunn.[308]

The history of the simply supported pipe problem is summarized in Table 6. Following the earlier analysis of Niordson,[309] Benjamin[114] made some theoretical and experimental observations on the behaviour of simply supported pipes during his major analysis of the cantilevered articulated pipe. Thurman and Mote[310] in 1969 made a study of the non-linear large vibrations of a pin-ended pipe, but did not discuss the static post-buckling response.

Paidoussis and Issid[307] made a major theoretical analysis together with a comprehensive review of the earlier literature in 1974. They studied the linear

Table 6 A summary of research work on the divergence of a simply supported pipe conveying fluid

		Continuous or articulated pipe	Includes experiment	Linear or non-linear	Includes damping
1954	Niordson	Continuous	No	Linear	No
1961	Benjamin	Articulated	Yes	Linear	No
1969	Thurman and Mote	Continuous	No	Non-linear vibration	No
1974	Paidoussis and Issid	Continuous	No	Linear	Yes
1975	Huseyin and Plaut	Continuous	No	Linear	No
1975	Plaut and Huseyin	Continuous	No	Linear	No
1977	Holmes	Continuous	No	Non-linear Euler p-b	Yes
1981	Holmes	Continuous	No	Non-linear Euler p-b	Yes

We note also a discussion of energy sources by Weaver, 1974, and a study of pipes with clamped ends due to Paidoussis, 1975.

behaviour of simply supported pipes both with and without damping, and observed the fluttering behaviour after initial divergence that we have discussed. They also observed how damping destroys any temporary gyroscopic stabilization. Huseyin and Plaut[311,312] analysed the linear undamped behaviour of a pinned pipe using a Galerkin discretization.

Finally Holmes in two important non-linear papers[113,313] looked at the Euler post-buckling of this pipe. In the second paper he emphasized that in the presence of damping such a pipe cannot exhibit sustained flutter in a limit cycle, due to the lack of a non-conservative energy source. Indeed the flutter with *increasing amplitude* that can be observed above the second Euler buckling load is driven by the system rolling *off an energy maximum*, as he observes: here the workless coriolis forces account for the *oscillatory* nature of the 'rolling-off' motion. This energy argument should be emphasized, because there has been a good deal of confusion about the energy source of the high-load fluttering behaviour.[314]

Before proceeding to the cantilevered pipe, we should note that a pipe clamped at both ends which behaves in a manner similar to that of the simply supported pipe[305] has been studied by Paidoussis[315] in 1975.

Table 7   A summary of research work on the flutter of a cantilevered pipe conveying fluid

		Flow in continuous or articulated pipe; or 2 links and follower-force	Includes experiment	Linear or non-linear	Includes damping
1961	Benjamin	Articulated with continuous as limit	Yes	Linear	Yes from coriolis
1964	Herrmann and Bungay	Follower-force	No	Linear	No
1965	Herrmann and Jong	Follower-force	No	Linear	Yes
1966	Herrmann and Nemat-Nasser	Follower-force	No	Linear	Yes
1966	Gregory and Paidoussis	Continuous	Yes	Linear	Yes plus coriolis
1967	Roorda and Nemat-Nasser	Follower-force	No	Non-linear +, damping	Yes
1970	Paidoussis	Continuous	Yes	Linear	Yes
1970	Paidoussis and Deksnis	Articulated and continuous	Yes	Linear	Yes from coriolis
1972	Burgess and Levinson	Follower-force	No	Non-linear +/−, stiffns	No
1974	Paidoussis and Issid	Continuous	No p. 281: Exp.	Linear limit cycles	Yes
1974	Bohn and Herrmann	Articulated	Yes with non-linear	Linear discussion	Yes
1976	Bishop and Fawzy	Continuous via classical modes	Yes	Linear	Yes
1977	Rousselet and Herrmann	Articulated	Short correlation	Non-linear +/−, mass	Yes plus coriolis

Plus or minus in the non-linear analyses denotes sign of the post-critical curvature, depending on the following system parameters.

The history of the fluttering cantilevered pipe is summarized in Table 7, starting with the classic paper of Benjamin[114] in 1961. He studied articulated pipes, both theoretically and experimentally, discussing also the continuous pipe in the limit as the number of links becomes large. In his linear theoretical work he derived the equation of motion using Lagrangian and Hamiltonian methods and observed that the coriolis forces provide a damping mechanism for a cantilevered pipe: for this reason it is not essential to introduce *structural* damping. Making no appeal to any linearization, he argued the irrelevance of fluid drag and he discussed the equivalent end compressive loads for both cantilevered and simply supported tubes. A cyclic energy expression was derived, as was the differential equation for a continuous pipe.

The next three papers by Herrmann and co-workers were concerned with the theoretical linear flutter of a two-link model subjected just to a follower force at the tip. Since no *coriolis* terms were included these papers (and the others classified in the table as 'follower-force') do not contribute directly to the fluid-loaded pipe since they only model the equivalent end compression associated with the *centrifugal* forces. Herrmann and Bungay[316] in 1964 looked at the effect of varying the proportion of conservative to non-conservative loading using a coefficient of lag, α, but with no damping in the model. Herrmann and Jong[143] in 1965 looked at the finite destabilization of infinitesimal damping with $\alpha = 0$ that we have summarized in Figure 33 of Chapter 1: they extended this a year later to the case of $\alpha \neq 0$. Herrmann and Nemat-Nasser[317] looked at the same damped model with $\alpha = 0$ by an energy method in 1966.

A significant *non-linear* study of the two-link follower-force model using an energy balance method due to Roorda and Nemat-Nasser[116] in 1967 showed *stable* post-flutter behaviour in the normal case, the positive post-critical curvature depending on the assumed damping ratio.

Meanwhile Paidoussis and co-workers in a series of major papers[119,307,318,319] between 1966 and 1974 looked at the theoretical and experimental behaviour of the continuous cantilevered pipe conveying fluid. In one of these, Paidoussis and Deksnis[119] drew attention to the fact that the theory for articulated pipes does not correlate well with the theory for continuous pipes even when the number of links becomes large. The theoretical work of these four papers was linear, but in the paper by Paidoussis and Issid[307] attention was briefly drawn to experimental evidence for non-linear limit cycles.

The non-linear analysis of the follower-force system due to Burgess and Levinson[320] in 1972 is limited by the fact that they include no damping, an infinitesimal amount of which can as we have seen make a finite change in the critical flutter load. They found positive and negative post-critical curvature depending on the non-linear stiffness ratios employed.

Bohn and Herrmann[321] in 1974 made a theoretical and experimental study of a two-link articulated pipe carrying a fluid flow. Damping was included in the linear theory, and some *non-linear* experimental evidence was discussed. This was followed by a big theoretical and experimental contribution from Bishop and

Fawzy[118] in 1976 who looked at the linear vibrations and stability of a hanging tube with a nozzle, both with and without harmonic forcing.

The final paper by Rousselet and Herrmann[117] in 1977 gives a non-linear theoretical analysis of a two-link articulated pipe conveying fluid, including damping. Stable and unstable points of dynamic bifurcation are observed, the sign of the post-critical curvature depending on the mass ratios, and a short correlation with experimental evidence is made. In this work the pressure head, rather than the flow velocity, is the prescribed quantity. This seems to be the definitive non-linear contribution to the pipe-flow problem at the present time, predicting either super- or sub-critical dynamic bifurcations, depending on the precise mass ratios of the articulated model.

A promising energy method that allows for a post-buckling analysis of non-conservative autonomous and non-autonomous systems is due to Leipholz.[322]

9.4 Resonance sensitivity of a dynamic Hopf bifurcation

We shall look, in this final section of our chapter on non-conservative problems, at a *non-autonomous* problem involving the periodic forcing of a non-linear oscillator. The forced oscillation of a non-linear system has in fact been an important problem in electrical circuit theory for many years, following the pioneering work of Van der Pol.[323,324] Recently some of the bifurcations involved have been examined by Holmes and Rand.[124,131]

We give here a discussion of a particularly simple non-linear oscillator which, when unforced, exhibits a dynamic bifurcation of the general Hopf type as its linear damping becomes negative: its response is therefore similar to a galloping structure. Such a structure might be resonated by periodic vortex shedding, which gives the motivation to our present study of the oscillator's forced response. This is based on a current paper by Thompson and Lunn.[325]

Now the galloping of a bluff elastically supported body in a steady fluid flow can be of the unstable dynamic type, and we could perhaps therefore expect some form of imperfection sensitivity as in the static case. It is known, however, that such a dynamic Hopf bifurcation is normally topologically stable under the operation of a single control parameter: this means that the topological form of the Hopf bifurcation cannot be rounded off or destroyed by the variation of a simple second control parameter, as is the case of the static cusp singularity.

However, since the dynamic instabilities are associated with a well-defined and non-zero circular frequency we might expect their failure 'loads' to be sensitive to resonant periodic forcing. Such a periodic forcing would not normally be contemplated in the assessment of topological stability, but could naturally arise in a real physical problem.

In fact we show here that a periodic forcing does indeed round off the Hopf bifurcation and gives rise to a two-thirds power-law sensitivity analogous to the static cusp. For a particular non-linear equation of motion considered the correspondence with the static bifurcation is indeed very precise, allowing the explicit use of standard static diagrams and formulae for the dynamic bifurcation.

Let us consider a one-degree-of-freedom non-linear oscillator with the equation of motion

$$\ddot{x} + b\dot{x} + cx + N(x, \dot{x}) = f \sin wt$$

where x is the displacement, b is the damping, c is the stiffness, and N represents non-linear terms: a dot denotes differentiation with respect to the time t. The forcing term on the right-hand side has amplitude f and the same circular frequency given by $w^2 = c$ as the corresponding undamped linear oscillator.

We assume now that as in unimodal galloping the net damping b decreases with the fluid velocity, a measure of the latter supplying us with a primary control parameter Λ. A dynamic instability is next assumed to arise as b, which incorporates positive structural damping and negative flow-induced damping, passes through zero at $\Lambda = \Lambda^C$, and we model this by writing

$$b = \Lambda^C - \Lambda.$$

With appropriate non-linearities $N(x, \dot{x})$ and $f = 0$ we shall now have an unstable Hopf bifurcation in which the limit cycle amplitude A is a positive constant k times the square root of the load decrement b,

$$A = kb^{1/2}, \qquad A^2 = k^2 b,$$

as shown in Figure 119. The superposition of some forcing, $f \neq 0$, will clearly trigger this instability at a value of Λ less than Λ^C and we can first *estimate* this by considering when the forced oscillation of *linear* resonance has the amplitude of the unstable limit cycle.

Now the amplitude of linear resonance, obtained by setting $N = 0$, is given for

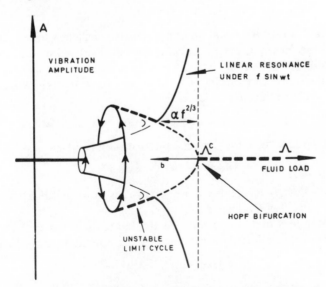

Figure 119   Heuristic demonstration of how a dynamic Hopf bifurcation might be sensitive to periodic forcing

light damping by the well-known result

$$A = \frac{f}{wb}$$

so for fixed f we have the asymptotically rising curve of Figure 119. This intersects the trace of limit cycles when

$$kb^{1/2} = \frac{f}{wb}$$

so we have the *result of our heuristic study*,

$$b = \left(\frac{f}{kw}\right)^{2/3}.$$

Thus since k and w are constants and $b = \Lambda^C - \Lambda$, we have

Reduction in load $\propto$ (forcing strength)$^{2/3}$

indicating a two-thirds power-law cusped resonance sensitivity.

The *heuristic* study suggests that we have a cusped sensitivity, and we make now an analysis of a *particular* non-linear oscillator that allows a *closed-form solution* for its steady vibrations.

Suppose that in our original equation we have the non-linear terms

$$N = D(x^2 + \dot{x}^2)\dot{x}$$

where D is a constant, and let us look for a solution in the form of a steady periodic response

$$x = A \cos wt.$$

Substituting this into the equation of motion the coefficient of $\cos wt$ vanishes because $w^2 = c$. Now the non-linear equation will clearly admit a simple closed-form solution if $w = 1$ so that $x^2 + \dot{x}^2 = A^2$, and we now assume this to be the case. The equation of motion containing now only $\sin wt$ terms is then completely satisfied if we have the *cubic relation*

$$DA^3 + bA + f = 0.$$

With $f = 0$ we now have the dynamic bifurcation formulae

$$A = 0 \qquad \text{or} \qquad A^2 = k^2 b$$

where $k^2 = -D^{-1}$, D being taken as negative to ensure an *unstable* bifurcation, as shown in Figure 8 of Chapter 1.

Non-zero values of f now round off the bifurcation *exactly* as in a *static* unstable-symmetric bifurcation governed by the potential energy function

$$V = \tfrac{1}{4}Dx^4 - \tfrac{1}{2}(\Lambda - \Lambda^C)x^2 + \varepsilon x$$

for which the equilibrium equation is

$$\frac{\partial V}{\partial x} = Dx^3 - (\Lambda - \Lambda^C)x + \varepsilon = 0.$$

This is identical to the above cubic relation with $A = x, b = -(\Lambda - \Lambda^C)$, as previously defined, and the forcing amplitude f equal to the imperfection parameter of the static theory, ε.

So we can now draw the well-known bifurcation diagram of Figure 17 of Chapter 1 with now a dynamic Hopf bifurcation at C and a *dynamic fold* for 'imperfect' systems. This dynamic fold involves the coalescence and vanishing of a stable limit cycle with an unstable limit cycle, and is the obvious dynamic analogy of the static fold or limit point. Using the general result of the static bifurcation theory (equation 8.3 on page 187 of our earlier monograph[36]), the resonance sensitivity is given by

$$\Lambda - \Lambda^C = 3D^{1/3}\left(\frac{f}{2}\right)^{2/3}$$

Comparison with the result of our heuristic study shows that the heuristic study obtained the correct *form* of solution but with the wrong *numerical* coefficient: this is not at all surprising.

Some numerical results for the steady-state solutions are shown in Figure 120. Here for a forcing amplitude $f = 0.1$ there are three steady-state oscillations at P, S, and C when $b = 0.5$: of these P is stable while the other two are unstable.

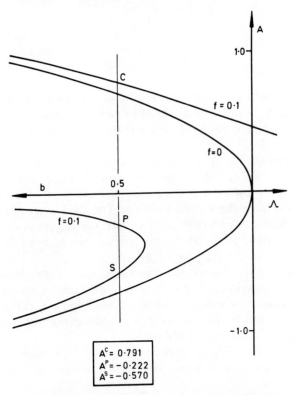

$A^C = 0.791$
$A^P = -0.222$
$A^S = -0.570$

Figure 120 Some computed curves of amplitude A against the control parameter Λ (or b)

182

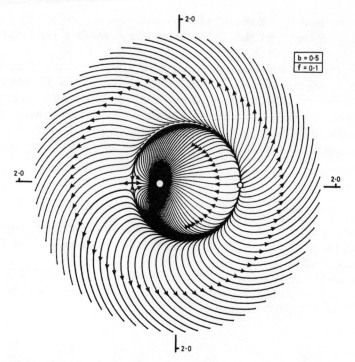

Figure 121 Two-dimensional phase trajectories for the smoothed
variational equation

Keeping the 'imperfection' f constant at 0.1 and decreasing b we see that P and S will merge and disappear at a critical value of b, leaving only the unstable vibration corresponding to C. We have here taken the coefficient D to be minus one.

To see more clearly the transitions between the steady states, we need to study the *transient* motions, and this is most conveniently done by assuming slowly varying amplitudes and plotting the smoothed trajectories of the autonomous variational equation in the Van der Pol plane. This gives us the phase portrait of Figure 121, which summarizes in polar coordinates the phase of the response relative to the forcing term and the amplitude of the response: details can be found in the aforementioned paper.[325]

In this picture the three steady states appear as equilibrium points, the central sink being the stable solution P. If we were now to decrease b, at constant $f = 0.1$, this sink would move towards the left-hand saddle S and at the critical value of b we would have a saddle-node coalescence. Thereafter the right-hand source C would be the only remaining steady state.

We see that at this saddle-node bifurcation the finite catchment area of P is instantaneously lost, all trajectories henceforth flowing from C out to infinity.

CHAPTER 10

Orbital Stability and the Attitude Control of Spacecraft

In this chapter we introduce first some ideas on the stability of motion,[326] including the concept of orbital as opposed to Liapunov stability. This is used to discuss the orbit of a small satellite attracted gravitationally by a massive body such as the earth or sun.

There follows a discussion of the passive nutation damping of a spinning spacecraft relevant to its attitude control, based on a recent paper by Kane and Levinson.[327] In an earlier work Kane and Barba[328] studied the effects of energy dissipation on a spinning satellite composed of two elastically connected rigid bodies: this work relates to one of the basic problems of space mechanics, that of the symmetric spinning satellite in a circular orbit. A year later, in 1967, Robe and Kane[329] made a comprehensive stability analysis of such an elastic satellite. They showed that certain vehicle configurations which are predicted to be stable when analysed as if the body were rigid must be classified as unstable when flexibility is taken into account.

10.1 Stability of motion of an orbiting particle

As a first example of the stability of motion, consider the planar motion of a mass m freely circling a fixed point C to which it is tied by a light inextensible string of length R. This system, shown in Figure 122, has only the single degree of freedom θ, so it will have a *two-dimensional* phase space generated by $(\theta, \dot{\theta})$. Since θ is cyclic, the phase space is best drawn on the surface of a cylinder as shown.

With no damping, the tension in the string is the only force on the particle, so all motions of the system have constant $\dot{\theta}$. The phase trajectories are thus circles as shown, the *phase velocity* of the circles increasing with $\dot{\theta}$ because of the reduction in the periodic time.

We consider now the stability of the *fundamental motion F*, in relation to a *perturbed motion P* which starts at $\theta = 0$ with a slightly larger value of $\dot{\theta}$. Because of the slightly different phase velocity, the representative points, P and F, will drawn steadily apart, until after some considerable time they will be widely separated: on a very long time scale, they will, for example, be often completely out of phase. So if we watch the *movement* of the phase *points* and make comparisons at various specific *times* we conclude that the fundamental motion F is *unstable in the sense of Liapunov*.

183

184

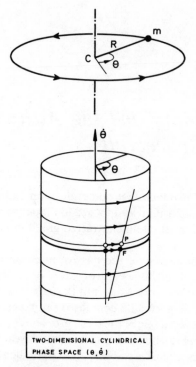

Figure 122 Two-dimensional cylindrical phase space for a mass circling on a light string

However, viewed in an overall *geometric* fashion and ignoring time, we see that the total phase *trajectory* of P is everywhere close to the trajectory of F. We therefore say that the motion F is *orbitally stable*.

As a second example in the stability of motion, we consider a satellite of mass m in orbit around a *fixed* mass M to which it is attracted by the inverse-square-law gravitation, as shown in Figure 123. Describing the position of the moving mass m by the polar coordinates (r, θ) we have the radial and angular equations of motion.

$$\ddot{r} - r\dot{\theta}^2 + \frac{GM}{r^2} = 0,$$

$$r\ddot{\theta} + 2\dot{r}\dot{\theta} = 0,$$

in which we have cancelled m throughout. These can be written down immediately from Newton's laws, or derived easily from the Lagrange equations. Here a dot denotes differentiation with respect to time and G is the gravitational constant.

The angular equation can be immediately integrated to give

$$r^2\dot{\theta} = p = \text{constant}$$

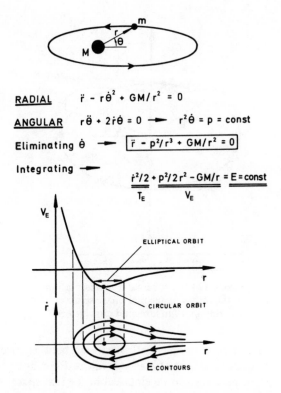

Figure 123 A satellite in orbit around a fixed mass *M*, showing the phase behaviour of an equivalent oscillator

which expresses the conservation of angular momentum. Eliminating $\dot{\theta}$ we now have the single equation for *r*,

$$\ddot{r} - \frac{p^2}{r^3} + \frac{GM}{r^2} = 0$$

which represents the equation of motion of a non-linear, one-degree-of-freedom oscillator. Integrating this once gives

$$\frac{\dot{r}^2}{2} + \frac{p^2}{2r^2} - \frac{GM}{r} = E = \text{constant}$$

which represents the conservation of total energy *E*. Here $\dot{r}^2/2$ is the effective kinetic energy of the oscillator, T_E, while the remainder of the left-hand side is the effective potential energy of the oscillator, V_E.

For a given *p* the form of V_E is shown in Figure 123 and in the two-dimensional phase space of the oscillator, $(r, \dot{r})$, contours of constant *E* give the phase trajectories shown. Here we identify one circular orbit (for the given *p*) with

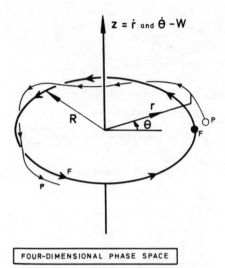

Figure 124 A schematic representation
of the four-dimensional phase space of a
satellite in a fundamental circular orbit

constant r, a family of closed orbits with r varying between two finite limits, together with unbounded motions at higher E values. Expressing r in terms of θ, the closed orbits can be shown to be elliptical in the real space.

We consider now the stability of a circular orbit of radius R for which

$$r = R, \qquad \dot{r} = 0, \qquad \theta \propto t, \qquad \dot{\theta} = W = \text{constant}.$$

To discuss its stability thoroughly we must examine all adjacent motions in the four-dimensional phase space $(r, \dot{r}, \theta, \dot{\theta})$ and as a schematic representation of this we draw r and θ as in the real space, allowing $\dot{r}$ and $\dot{\theta}$ to *share* the z direction perpendicular to the (r, θ) plane. This is shown in Figure 124.

The fundamental circular motion F is now the heavy circle drawn. A slightly perturbed motion P will in general have a slightly different value of p and therefore a slightly different curve for V_{E} and a slightly different value E. It follows from our oscillator discussion that P will be either circular or closed with a *small circuit* in the $(r, \dot{r})$ diagram.

It is clear therefore that for all time the adjacent perturbed motion will have an r close to R, an $\dot{r}$ close to zero, and a $\dot{\theta} = p/r^2$ close to W. So the trajectory of P will always be geometrically close to the trajectory of F, and we declare the fundamental circular motion to be *orbitally stable*.

However, the fundamental motion is clearly *unstable* in the sense of Liapunov because the adjacent trajectories will in general have different periodic times. For example the equation of a circular orbit is

$$m\omega^2 r = \frac{GmM}{r^2}$$

so its periodic time is

$$T = \frac{2\pi}{\omega} = 2\pi \sqrt{\frac{r^3}{GM}}$$

which is a function of r. So the Liapunov instability can be demonstated by discussing just the special perturbation that carries the fundamental motion into an adjacent circular one.

10.2 Passive nutation damping of a spinning satellite

This section, on the physical significance of spacecraft instabilities, was drafted by Professor Tom R. Kane of Stanford University, based on a recent article of his with Levinson[327] and with the permission of the American Institute of Aeronautics and Astronautics.

Certain artificial satellites of the Earth consist primarily of a single rigid body that is intended to move in such a way that one of its central principal axes of inertia remains at all times normal to the plane of the orbit traversed by the mass centre of the body. To maintain this state of motion, it can be advantageous to permit the satellite to spin, that is, to have an angular velocity normal to the orbit plane and to equip the satellite with a nutation damper, that is, with a device that dissipates energy whenever the satellite moves in a manner other than that intended and thus brings about a resumption of the desired motion. However, spin and damping can have also adverse effects. Unless certain system parameters are chosen properly, it can occur that the slightest disturbance causes the satellite to tumble. Mathematical stability theory permits one to identify dangerous values of the parameters, and solutions of the differential equations governing the motion of a satellite reveal the physical significance of instabilities. These ideas are illustrated by means of an example in what follows.

In Figure 125 B designates a rigid body carrying a particle P that is attached to a spring S and a dashpot D. Y_1, Y_2, Y_3 are principal axes of inertia of B for the mass centre B^* of B, hereafter called central principal axes. P is constrained to move on a line parallel to Y_1 and S is presumed to be undeformed when P lies on Y_2. In the absence of external forces, this system can perform a motion of *simple spin*. That is, it can move in such a way that P remains on Y_2 and the orientation of Y_1 in an astronomical reference frame A remains fixed while the angular velocity of B in A has a constant magnitude Ω and is permanently parallel to Y_1. If such a motion is disturbed at some instant of time $t = 0$, then the orientation of Y_1 in A and the distance q between P and Y_2 generally vary with time for $t > 0$, and the simple spin under consideration is said to be unstable if one cannot keep both q and the departure of Y_1 from its original orientation arbitrarily small by making the disturbance sufficiently small.

Three conditions, violation of any one of which guarantees instability, can be formulated in terms of the spin speed Ω and parameters characterizing the body B and the nutation damper elements P and S. If we take B to be a uniform,

188

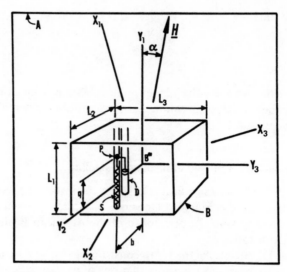

Figure 125   Schematic representation of a spacecraft

rectangular block having a mass density ρ and sides of lengths L_1, L_2, L_3 as drawn, let P have a mass v times that of B, and choose for S a linear spring with a spring constant σ, then a procedure described in an earlier paper[330] leads to the conclusion that the simple spin under consideration is unstable whenever at least one of u_1, u_2, u_3 is negative, these three quantities being defined as

$$u_1 = L_3 - L_1,$$
$$u_2 = (L_2^2 - L_1^2)(1 + v) + 12vb^2,$$
$$u_3 = \frac{\sigma(1 + v)}{v\rho L_1 L_2 L_3 \Omega^2} - \frac{12vb^2}{u_2}.$$

The two conditions $u_1 < 0$, $u_2 < 0$ may be termed 'non-maximum inertia' criteria, for satisfaction of either one guarantees that the angular velocity vector characterizing the simple spin is parallel to a line other than the axis of maximum central moment of inertia of the system formed by B and P. The condition $u_3 < 0$ is called a 'compliance' criterion, for it can be satisfied only when the spring stiffness σ is sufficiently small.

All three instability conditions can be *violated* simultaneously in only one way, indicated in Figure 126 by the letter N, standing for 'not unstable'; but one or more of the conditions can be satisfied in the five ways associated with the letters $U_1, \ldots, U_5$. Since u_3 involves u_2 in such a way that $u_2 < 0$ implies $u_3 \geq 0$, the two combinations $u_1 > 0$, $u_2 < 0$, $u_3 < 0$ and $u_1 < 0$, $u_2 < 0$, $u_3 < 0$ cannot occur, and are therefore omitted from the figure. In the case of U_1, one is dealing with a spacecraft for which the simple spin under consideration would not be unstable if P were rigidly attached to B; for, with $u_1 > 0$ and $u_2 > 0$, the spin axis is parallel to the axis of maximum central moment of inertia (when P lies on Y_2), and for a rigid

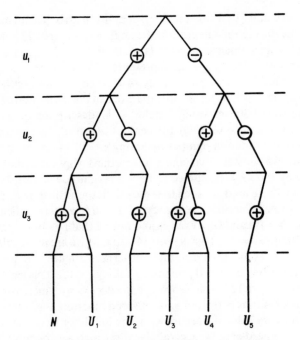

u_1

u_2

u_3

N $\quad U_1$ $\quad U_2$ $\quad U_3$ $\quad U_4$ $\quad U_5$

Figure 126 Instability conditions for the attitude of the craft

body this guarantees marginal stability. Hence U_1 represents destabilization attributable to the nutation damper. The possibility of this occurrence was pointed out by Pringle[331] and the subject is discussed also by Kane and Teixeira.[332] U_2, U_3, and U_4 all correspond to motions that would be unstable even if P were fixed on B, because the spin axis is parallel to a central principal axis that is neither the axis of maximum nor the axis of minimum central moment of inertia. For a rigid body, the associated instability is well known to manifest itself in the form of large excursions of Y_1, so we shall not pursue U_2, U_3, and U_4 further. By contrast, U_5 again involves a spacecraft in connection with which the mobility of P is of crucial importance, the spin axis here being parallel to the axis of minimum central moment of inertia, so that, if P were fixed on B, one would have marginal stability. Moreover, instability now cannot be blamed on excessive compliance, since $u_3 > 0$, but may be viewed as representing destabilization arising from energy dissipation.

After assigning numerical values to all system parameters, one may proceed as follows: write dynamical differential equations governing q and the Y_1, Y_2, Y_3 measure numbers, $\omega_1, \omega_2, \omega_3$, of the angular velocity ω of B in A; write kinematical differential equations involving $\omega_1, \omega_2, \omega_3$ and the elements of the direction cosine matrix $[C_{ij}]$ that relates Y_1, Y_2, Y_3 to axes X_1, X_2, X_3 fixed in A and initially coincident with Y_1, Y_2, Y_3 respectively; integrate all differential equations simultaneously, using the initial conditions $q = \dot{q} = 0$, $\omega_1 = \Omega$,

$\omega_2 = \varepsilon\Omega$, $\omega_3 = 0$, and $[C_{ij}] = I$, the 3×3 identity matrix; calculate for discrete instants of time during the integration the angle α of Figure 125 between Y_1 and the central angular momentum vector **H** of the system which remains fixed in A and plot α against t. Here $\omega_2 = \varepsilon\Omega$ represents a disturbance of a simple spin.

To set the stage for subsequent study of U_1 and U_5, we first consider N in Figure 126, taking $\rho = 2760\,\text{kg/m}^3$ (the mass density of aluminum), $L_1 = 1.200\,\text{m}$, $L_2 = 1.225\,\text{m}$, $L_3 = 1.300\,\text{m}$, which concludes the description of B. Next, we set $v = 0.01$, so that the mass of P is 1 per cent. of that of B, and then complete the specification of the nutation damper by letting $\sigma = 52.744\,\text{N/m}$ and $\delta = 105.487$ N s/m, where δ is a constant that characterizes the dashpot D in the sense that the force exerted by D on P is presumed to have a magnitude of $\delta|\dot{q}|$. The numerical values of σ and δ here used are such that the oscillator formed by P, S, and D has a natural (undamped) circular frequency of 1 rad/s and is critically damped. Finally, we set $b = 1\,\text{m}$ and $\Omega = 1\,\text{rad/s}$, and noting that by our equations we now have $u_1 = 0.100\,\text{m}$, $u_2 = 0.181\,\text{m}^2$, $u_3 = 0.348$, so that we are indeed dealing with a motion that violates all three instability criteria of Figure 126, we perform the numerical integration described previously, taking $\varepsilon = 0.1$. This leads to the curve labeled $N(.1)$ in Figure 127, which shows that α decays with increasing time t. The nutation damper is thus performing its intended function. It is interesting to note that this satisfactory behavior occurs when the oscillator has a natural frequency equal to the spin speed of the spacecraft. Next, we increase the compliance of S by decreasing σ to $26.372\,\text{N/m}$, one-half of its former value, but leave all other parameter values unchanged, which has the effect of making $u_3 = -0.157$ while u_1 and u_2 retain their former positive values. Hence we are dealing with U_1 in

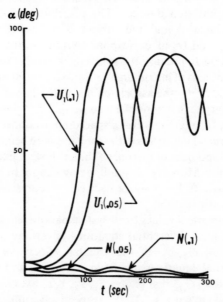

Figure 127 Stable and unstable maximum axis spins

Figure 126. The same disturbance used previously with $\varepsilon = 0.1$ now gives rise to the curve labeled $U_1(.1)$ in Figure 127, from which it is immediately apparent that the nutation 'damper' is, in fact, acting as a nutation generator. Now, one might conjecture that the large response here encountered is attributable to an excessively large initial disturbance. To see that this is not the case, one can change ε from 0.1 to 0.05 and then re-examine both N and U_1, which was done to produce the curves $N(.05)$ and $U_1(.05)$. The four curves in Figure 127 illustrate a fundamental proposition. When a motion is stable, reducing a disturbance has the effect of reducing departures from the nominal motion; but when a motion is unstable, reducing a disturbance merely defers such departures. Thus, $N(.05)$ lies below $N(.1)$ for all time whereas $U_1(.05)$ attains values comparable to the largest values of $U_1(.1)$, albeit more slowly.

An instability need not manifest itself so unequivocally as did the one just examined. For example, if the dimensions of B are changed to $L_1 = 0.500$ m, $L_2 = 1.200$ m, $L_3 = 3.185$ m, and the spring constant is changed to $\sigma = 0.527$ N/m, but ρ, ν, δ, b, and Ω are left unaltered (B has the same mass as heretofore, and the spring is one hundred times more compliant than in the preceding case), then $u_1 = 2.685$ m, $u_2 = 1.322$ m^2, $u_3 = -0.081$, so that one is still dealing with U_1 in Figure 126. But now, for the same disturbance as before with $\varepsilon = 0.1$, one obtains the rather more complex response depicted in Figure 128, one interesting feature of which is that α remains quite small for the first two minutes of the motion, giving the appearance of, at least, marginal stability, but then grows very rapidly. This plot illustrates a second fundamental fact. When a motion is of practical interest for only a limited time interval, instability of the motion

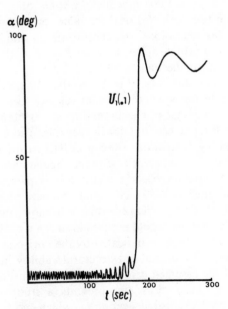

Figure 128 Potentially deceptive maximum axis spin

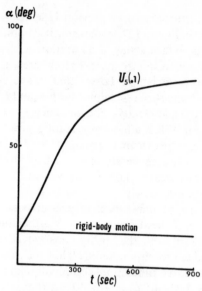

Figure 129   Unstable minimum axis spin, compared with a rigid-body motion

may be tolerable, because objectionably large departures from the nominal motion may occur only subsequent to the time of interest.

The instability associated with U_5 in Figure 126 arises, for example, if $L_1 = 2.000\,\text{m}$, $L_2 = 0.948\,\text{m}$, $L_3 = 1.008\,\text{m}$ while $\rho, v, \sigma, \delta, b$, and Ω have the values used earlier in connection with N. For these values of the system parameters, B has the same mass as heretofore; the central inertia ellipsoid of the system formed by B and P is a prolate spheroid when P is on Y_2 (the central principal moments of inertia have the values 894 and 2205 kg m^2), and the axis of revolution of this spheroid is parallel to Y_1, so that the nominal motion under consideration is one during which the angular velocity vector of B is parallel to the axis of minimum moment of inertia; and $u_1 = -0.992\,\text{m}$, $u_2 = -3.012\,\text{m}^2$, $u_3 = 1.050$. With $\varepsilon = 0.1$, one here obtains the curve labeled $U_5(.1)$ in Figure 129, which shows that energy dissipation now causes B to go into a 'flat spin'; that is, the Y_1 axis approaches perpendicularity with the angular momentum vector. The claim that energy dissipation plays a major role is particularly strong here because, if P were fastened to Y_2, which would eliminate energy dissipation, the rigid body formed by B and P would perform a well-known motion of precession accompanied by spin, and this would proceed in such a way that α remains at all times equal to its initial value, as indicated by the horizontal line.

A discussion of Liapunov, orbital and structural stability, including a survey of the type and size of the assumed disturbances (deterministic and stochastic) is made by Mazzilli.[333] This thesis also presents a study of the parametric instabilities of a forced extensible pendulum, which have similarities to the Mathieu instabilities of tension-leg oil platforms.

CHAPTER 11

Dynamical Field Theories for Neural Activity in the Brain

We choose to end this survey of instabilities and catastrophes by looking at the important and exciting modelling of man's brain. Here the relationships between the apparently simple activity of the individual neurons and the high levels of organization associated with thought and consciousness are only just beginning to be explored. This account can be seen as a natural extension to our discussion in Chapter 5 of the self-organization of chemical and biochemical systems.

To model the neural tissue of the brain, Wilson and Cowan[334] derive the dynamical equations of populations of excitatory and inhibitory neurons. For spatially localized populations, coupled non-linear differential equations are obtained, and have been studied using phase-plane methods and numerical analysis. *Folds* in the steady state solutions are found to generate multiple hysteresis phenomena, while limit cycles (modelling brain rhythms) are observed in which the frequency of oscillation is found to be a monotonic function of stimulus intensity. An extension of this work to include a spatial distribution of the tissue is summarized by Wilson.[335]

11.1 The brain and central nervous system

The brain can be highly idealized as a network of *neurons* connected in a random manner by *synapses*. When a neuron *fires* the stimulus is transmitted through the synaptic connections to adjacent neurons, which may then be induced to fire after the synaptic *delay*.

The neuron population can be divided into *excitatory* neurons, which give out a positive stimulus when they fire, and *inhibitory* neurons, which give out a negative stimulus. A neuron will fire when the sum of the received stimuli exceeds a certain *threshold* value: and having once fired it remains inactive for a certain *refractory* period, even if it receives stimulus above its threshold.

Such a discrete *neural net* can be readily modelled on a digital computer, and waves of firing activity have been observed in computer simulations as discussed by Anninos.[336]

In an attempt to model the *higher* functions of the brain and central nervous system it seems appropriate however to discuss not the behaviour of a discrete system of randomly connected neurons but, rather, a *continuum* of neuron populations in which the randomness is smeared out to give a deterministic behaviour. This is exactly analogous to the continuum field theories of solid and fluid mechanics. In the theory of elasticity, for example, the random vibrations of the atoms play no part, owing to the introduction of bulk macroscopic moduli. Similarly a fluid observed at the molecular level reveals random thermal Brownian motions, although seen macroscopically the fluid may be exhibiting smooth laminar flow. This deterministic smearing approach is validated in brain studies by the *local redundancy* of the neurons in a small volume of cortical tissue.

The work of Wilson and Cowan[334] is distinguished from much analogous activity by their important inclusion of both excitatory *and* inhibitory neurons.

11.2 Mechanics of excitation and inhibition

The model of Wilson and Cowan introduces as two fundamental variables $E(t)$, the proportion of excitatory cells firing per unit time, and $I(t)$ the proportion of inhibitory cells firing per unit time. Notice that these are both *rates* of firing and, to eliminate some short-term effects, we must strictly regard them as moving time averages over some small period of time.

It is assumed that E and I at time $(t + \tau)$ after a *delay* τ will be equal to the proportion of cells which are *sensitive* and which also receive at least *threshold* excitation.

Non-sensitive cells are those that, having recently fired, cannot fire again for their *refractory* period. If the absolute refractory period is r, the proportion of sensitive excitatory cells can be approximated as

$$E_s = 1 - r_e E$$

with a similar expression for I_s. Notice that the refractory period for E is written as r_e which might be different from r_i.

Now the expected proportions of the sub-populations receiving at least threshold excitation per unit time will be a mathematical *function* of E and I which we write for the excitatory cells as

$$\mathscr{S}_e(x) = \mathscr{S}_e[c_e E - g_e I + P(t)]$$

and for the inhibitory cells as

$$\mathscr{S}_i(x) = \mathscr{S}_i[c_i E - g_i I + Q(t)].$$

Here the coefficients are constants representing the average number of synapses per cell, and $P(t)$ and $Q(t)$ are *external* excitations.

The *response functions* $\mathscr{S}(x)$ will depend on the probability distribution of neural thresholds. It is argued that they will have the *sigmoidal* shape of an integral sign, rising monotonically with x from zero and becoming asymptotic to a value equal to or near to unity as x tends to infinity. In the analytical work they are taken as

$$\mathscr{S}(x) = \frac{1}{1 + \exp[-a(x - \theta)]} - \frac{1}{1 + \exp(a\theta)}$$

with different values of the constants a and θ for the two types of neuron.

Now if the probability of a cell being sensitive is independent of the probability that it is currently excited above its threshold, we can multiply our probabilities to get, with some time coarse *graining* assumptions[334]

$$E(t + \tau_e) = (k_e - r_e E)\mathcal{S}_e[c_e E - g_e I + P(t)]$$

$$I(t + \tau_i) = (k_i - r_i I)\mathcal{S}_i[c_i E - g_i I + Q(t)].$$

Here k_e and k_i replace unity in our earlier expressions for E_s and I_s: they are in fact very close to unity, being defined as

$$k = \mathcal{S}(\infty).$$

They are part of a small adjustment to make $E = I = 0$ a stable *resting* state under zero external excitation.

If we now write the Taylor approximation

$$E(t + \tau_e) = E(t) + \frac{dE}{dt}\tau_e$$

and likewise for I, we have our final differential equations

$$\tau_e \frac{dE}{dt} = -E + (k_e - r_e E)\mathcal{S}_e[c_e E - g_e I + P(t)]$$

$$\tau_i \frac{dI}{dt} = -I + (k_i - r_i I)\mathcal{S}_i[c_i E - g_i I + Q(t)].$$

In our particular study of these we shall put the external stimulation of the inhibitory neurons to zero ($Q = 0$) and replace $P(t)$ by a constant P, which can then be regarded as a *control parameter*.

11.3 Folds and multiple hysteresis in the steady states

With $Q = 0$ and a time-independent P as our control, the steady states given by

$$\frac{dE}{dt} = \frac{dI}{dt} = 0$$

exhibit multiple folding and hysteresis for certain values of the coefficients as shown for example in Figure 130. This demonstration of hysteresis is most important because hysteresis has been suggested as a physiological basis for short-term memory: and there is at least one experimental verification of hysteresis within the central nervous system.

11.4 Temporal oscillations in a limit cycle

It is shown by Wilson and Cowan[334] that the present model can exhibit a *damped* oscillatory response to *impulsive* external stimulation, as indeed could be demanded of a satisfactory modelling.

196

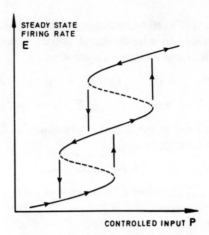

Figure 130 Typical multiple folds and hysteresis in the steady state firing rate E as a function of the controlled input P with $Q = 0$

$$E + \tau \frac{dE}{dt} = (k - rE)\, \mathcal{f}(cE - gI + P)\Big|_e$$

$$I + \tau \frac{dI}{dt} = (k - rI)\, \mathcal{f}(cE - gI + Q)\Big|_i$$

$$\mathcal{f}(x) = \frac{1}{1 + \exp[-a(x - \theta)]} - \frac{1}{1 + \exp(a\theta)}$$

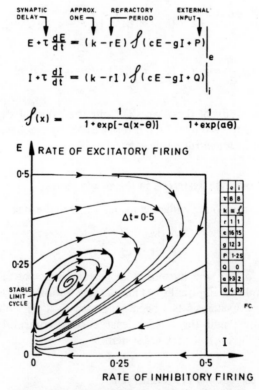

Figure 131 A stable limit cycle in the (E, I) phase space with the set of coefficients displayed. The limit cycle represents a steady oscillation in the two firing rates

Moreover, with Q equal to zero and P equal to a certain constant value, the model can, with an appropriately chosen set of coefficients exhibit a stable limit cycle as shown in the two-dimensional (E, I) phase space of Figure 131. These limit cycles arising from a realistic neural model provide a concrete physiological base for the study of electroencephalogram (EEG) rhythms, such as the important alpha rhythms.

11.5 Pattern formation in the space of a neural field

The next vital step in the study of neural tissue is to examine the behaviour of our model when it is distributed in space to form a *neural field*. We have already seen how this can be done for the Brusselator model chemical reaction in Chapter 5 where we examined the spontaneous emergence of order in chemical and biochemical fields. Work on the development of the necessary neural field equations is reported by Wilson[335] in a series of lectures edited by H. Haken, the creator of the new discipline of Synergetics. Similar work is due to Amari.[337]

The field equations allow us to examine the spatial interactions within sheets of neural tissue characteristic of the cortex of the brain: and important detailed effects still to be investigated are those due to the geometry, size and boundary conditions of the neural field. Already many features of *pattern formation* and *learning* in neural pools and fields have been investigated. Spatially-localized limit cycles are demonstrated by Wilson,[335] and the appearance of a cusp catastrophe is discussed by Amari.[337]

APPENDIX

Some Biographies in the History of Stability

Leonard Euler, 1707 – 1783

Historically we could well begin the story of stability with the Swiss mathematician Leonard Euler who determined the bifurcating equilibrium configurations of a compressed elastic column in an appendix to his major work on the calculus of variations[2] published in 1744. Euler was one of the most prolific researchers in the history of mathematics, and was called by his contemporaries 'analysis incarnate'. He opened his career in the year of Newton's death and brought the full power of the calculus to bear on mechanics in his analytical treatise of 1736.

Euler had the enviable ability to work anywhere under any conditions, and typically dashed off a paper in the half-hour before dinner. He was aided by a prodigious memory and powers of mental arithmetic, which served him well when he was afflicted with total blindness: his productivity actually increased, and he performed the complex analysis of his three-body lunar theory entirely in his head.

His mind remained clear and powerful up to the second of his death by a stroke in 1783.

Joseph-Louis Lagrange, 1736 – 1813

A young friend of Euler was the great French mathematician Joseph-Louis Lagrange, addressed by Napoleon as the 'lofty pyramid of the mathematical sciences'. His rich father had squandered the bulk of his possessions in reckless speculation, but Lagrange understanding the web of life, philosophized 'If I had inherited a fortune I should probably not have cast my lot with mathematics'. As it turned out, Lagrange was appointed professor of mathematics at the age of 16, teaching students all older than himself, and went on to a most brilliant career.

He perfected the analytical rather than the geometrical approach to mechanics and pointed out in the preface to his master-piece *Méchanique Analytique*[3] that 'no diagrams will be found in this work'. Nevertheless he observed that mechanics was based on the analytical geometry of four-dimensional space–time, a viewpoint that reached maturity with Albert Einstein.

The Lagrange equations continue to play a vital role in mechanics, bringing a greater generality of approach than is possible with the vectorial methods of

Newton: and they lead him quite naturally to the concept of a minimum of the total potential energy for the stability of a conservative system.

His book appeared after some delays in 1788, one year before the fall of the Bastille. But the first copy found Lagrange in a state of chronic depression and is reported to have lain unopened on his desk for two years.

During the period of the French Revolution he was a leading originator of the metric system of weights and measures and saved us the disaster of a proposed base 12 system! In his old age his revived mathematical enthusiasms were channelled into the second edition of his *Méchanique Analytique*.

Henri Poincaré, 1854–1912

Henri Poincaré was one of the giants of mathematics and was perhaps the last man to embrace in his own work the whole of the mathematics of his day. It is generally acknowledged that the exponential growth of mathematics has precluded any such embrace today. He published over five hundred scientific papers on new topics in mathematics, together with more than thirty books.

He was a deeply philosophic man, but physically awkward with poor muscular control as a result of which he once scored zero in a drawing examination. The horror of accidentally shooting a bird in his youth affected him deeply for years after the event.

Like Euler before him, he had a powerful 'visual' memory and made tracts of complex analysis entirely in his head on top of any disturbance. Despite, or maybe because of, this he was extremely absent-minded in his everyday life. He laid the foundations of modern bifurcation theory, which are sketched lucidly in an article on rotating liquid masses,[4] and initiated a continuing programme of qualitative global dynamics.

Poincaré was one of the first to recognize the revolutionary importance of Einstein's work, and he made a psychological study of mathematical creativity concluding 'Mathematical discoveries, small or great...are never born of spontaneous generation. They always presuppose a soil seeded with preliminary knowledge and well prepared by labour, both conscious and subconscious'.

References

1. I. Newton, *Mathematical Principles of Natural Philosophy*, 1686. Translated into English by Andrew Motte in 1729. Reprinted by the University of California Press, Berkeley, 1974.
2. L. Euler, *Methodus Inveniendi Lineas Curvas Maximi Minimive Proprietate Gaudentes* (Appendix, De curvis elasticis), Marcum Michaelem Bousquet, Lausanne and Geneva, 1744.
3. J. L. Lagrange, *Méchanique Analytique*, Courcier, Paris, 1788.
4. H. Poincaré, Sur l'equilibre d'une masse fluide animée d'un mouvement de rotation, *Acta. math.*, **7**, 259 (1885).
5. H. Poincaré, *Les Méthodes Nouvelles de la Mécanique Céleste*, Vols 1–3, Gauthier-Villars, Paris, 1892–1899.
6. H. Poincaré, *Oeuvres*, Gauthier-Villars, Paris, 1951.
7. A. Liapunov, *Problème Général de la Stabilité du Mouvement*, Kharkov, 1892. French translation in *Ann. fac. sci. univ. Toulouse*, **9**, 1907. Reproduced in *Ann. Math. Studies*, Vol. 17, Princeton University Press, Princeton, N. J., 1949.
8. A. A. Andronov and L. S. Pontryagin, Coarse systems, *Dokl. Akad. Nauk. SSSR*, **14**, 247 (1937); also in A. A. Andronov, Sobraniye trudov, *Izd. Akad. Nauk SSSR*, **1956**, 181 (1956).
9. R. Thom, *Structural Stability and Morphogenesis*. Translated from the French by D. H. Fowler, Benjamin, Reading, 1975.
10. E. C. Zeeman, *Catastrophe Theory: Selected Papers 1972–1977*, Addison Wesley, London, 1977.
11. T. Poston and I. Stewart, *Catastrophe Theory and its Applications*, Pitman, London, 1978.
12. S. Smale, On dynamical systems, *Bol. Soc. Mat. Mexicana*, **1960**, 195–198 (1960).
13. S. Smale, Differentiable dynamical systems, *Bull. Am. Math. Soc.*, **73**, 747 (1967)
14. V. I. Arnold, Small denominators and problems of stability of motion in classical and celestial mechanics, *Russian Mathematical Surveys*, **18** (6), 85 (1963).
15. V. I. Arnold, Lectures on bifurcations in versal families, *Russian Maths Surveys*, **27**, 54 (1972).
16. R. Abraham and J. E. Marsden, *Foundations of Mechanics*, 2nd ed., Benjamin, Reading, 1978.
17. W. T. Koiter, *On the Stability of Elastic Equilibrium,* Dissertation, Delft, Holland, 1945. (An English translation is now available as NASA, *Tech. Trans.*, **F 10**, 833, 1967.)
18. B. Budiansky, Theory of buckling and post-buckling behaviour of elastic structures, *Advances in Applied Mechanics*, Vol. 14, Academic Press, New York, 1974.
19. J. W. Hutchinson, Plastic buckling, *Advances in Applied Mechanics*, Vol. 14, Academic Press, New York, 1974.
20. H. Ziegler, *Principles of Structural Stability*, 2nd ed., Birkhauser Verlag, Basel, 1977.

21. G. Herrmann, Stability of equilibrium of elastic systems subjected to non-conservative forces, *Appl. Mech. Rev.*, **20**, 103 (1967).
22. H. H. E. Leipholz, *Stability Theory: An Introduction to the Stability of Dynamic Systems and Rigid Bodies*, Academic Press, New York, 1970.
23. H. H. E. Leipholz, *Six Lectures on Stability of Elastic Systems*, 2nd ed. (revised and enlarged), Solid Mechanics Division, University of Waterloo, Waterloo, 1974.
24. H. H. E. Leipholz, *Direct Variational Methods and Eigenvalue Problems in Engineering*, Noordhoff, Leyden, 1977.
25. H. H. E. Leipholz, *Stability of Elastic Systems*, Sijthoff and Noordhoff, Alphen, 1980.
26. J. M. T. Thompson, Basic principles in the general theory of elastic stability, *J. Mech. Phys. Solids*, **11**, 13 (1963).
27. J. Roorda, Stability of structures with small imperfections, *J. Engng Mech. Div. Am. Soc. civ. Engrs*, **91**, 87 (1965). See also, J. Roorda, *Buckling of Elastic Structures*, Special Publications Series, Solid Mechanics Division, University of Waterloo Press, Waterloo, 1980.
28. J. Roorda, The buckling behaviour of imperfect structural systems, *J. Mech. Phys. Solids*, **13**, 267 (1965).
29. J. Roorda, On the buckling of symmetric structural systems with first and second order imperfections, *Int. J. Solids Structures*, **4**, 1137 (1968).
30. J. M. T. Thompson, A general theory for the equilibrium and stability of discrete conservative systems, *Z. angew. Math. Phys.*, **20**, 797 (1969).
31. J. G. A. Croll and A. C. Walker, *Elements of Structural Stability*, Macmillan, London, 1972.
32. W. J. Supple (Ed.), *Structural Instability*, IPC Science and Technology Press, Guildford, 1973.
33. K. Huseyin, *Non-linear Theory of Elastic Stability*, Noordhoff, Leyden, 1974.
34. J. M. T. Thompson and G. W. Hunt, The instability of evolving systems. *Interdisciplinary Science Reviews*, **2**, 240 (1977).
35. K. Huseyin, *Vibrations and Stability of Multiple Parameter Systems*, Noordhoff, Alphen, 1978.
36. J. M. T. Thompson and G. W. Hunt, *A General Theory of Elastic Stability*, Wiley, London, 1973.
37. J. M. T. Thompson and G. W. Hunt, Static and Dynamic Instability Phenomena, Wiley, in press.
38. J. M. T. Thompson, Experiments in catastrophe, *Nature*, **254**, 392 (1975).
39. J. M. T. Thompson and G. W. Hunt, Towards a unified bifurcation theory, *J. Appl. Math. Phys. (ZAMP)*, **26**, 581 (1975).
40. G. Herrmann, Determinism and uncertainty in stability, in *Instability of Continuous Systems* (Ed. H. Leipholz), Springer, Berlin, 1971.
41. H. Ziegler, Trace effects in stability, in *Instability of Continuous Systems* (Ed. H. Leipholz), Springer, Berlin, 1971.
42. J. M. T. Thompson, Basic theorems of elastic stability, *Int. J. Engng. Sci.*, **8**, 307 (1970).
43. D. R. J. Chillingworth, *A Problem from Singularity Theory in Engineering*, Lecture to Symposium on Non-linear Mathematical Modelling, University of Southampton, August, 1976.
44. E. Hopf, Abzweigung einer periodischen Losung von einer stationaren Losung aines Differentialsystems, *Berichten der Math.-Phys. Klass der Sachlischen Akademie der Wissenschaften zu Leipzig*, **94**, 3 (1942).
45. J. Marsden and M. McCracken, *The Hopf Bifurcation and its Applications*, Springer Applied Maths Series, No. 19, Springer, Berlin, 1976.
46. J. M. T. Thompson, Catastrophe theory and its role in applied mechanics, *Proc*

fourteenth IUTAM Congress, Delft, August, 1976, North Holland, Amsterdam, 1976/7.
47. J. M. T. Thompson, Stability predictions through a succession of folds, *Phil. Trans. Roy. Soc. Lond.*, Ser. A, **292**, 1–23 (1979).
48. G. I. Taylor, Disintegration of water drops in an electric field, *Proc. Roy. Soc. Lond.*, Ser. A, **280**, 383 (1964).
49. E. C. Zeeman, Catastrophe theory, *Scientific American*, **234**, No. 4, 65 (1976).
50. T. B. Benjamin, Bifurcation phenomena in steady flows of a viscous fluid. I. Theory, II. Experiments, *Proc. Roy. Soc. Lond.*, Ser. A, **359**, 1–26 and 27–43, 1978.
51. D. Postle, *Catastrophe Theory*, Fontana, London, 1980.
52. J. G. A. Croll, Is catastrophe theory dangerous?, *New Scientist*, **70**, 630 (1976).
53. R. S. Zahler and H. J. Sussmann, Claims and accomplishments of applied catastrophe theory, *Nature*, **269**, 759 (1977). See also subsequent correspondence in **270**, 381 (1977) and **271**, 401 (1978).
54. H. J. Sussmann and R. S. Zahler, Catastrophe theory as applied to the social and biological sciences: a critique, *Synthese*, **37**, 117 (1978).
55. A. Woodcock and M. Davis, *Catastrophe Theory*, Dutton, New York, 1978.
56. D. H. Michael, M. E. O'Neill and J. C. Zuercher, The breakdown of electrical insulation in a plane layer of insulting fluid by electrocapillary instability, *J. Fluid Mech.*, **47**, 609 (1971).
57. D. H. Michael and M. E. O'Neill, Two-dimensional problems of electrohydrostatic stability, *Phil. Trans. Roy. Soc.*, Ser. A, **272** (1224), 331 (1972).
58. D. H. Michael and M. E. O'Neill, The bursting of a charged cylindrical film, *Proc. Roy. Soc. Lond.*, Ser. A, **328**, 529 (1972).
59. D. H. Michael, J. Norbury, and M. E. O'Neill, Electrohydrostatic instability in electrically stressed dielectric fluids, *J. Fluid Mech.*, **66**, 289 (1974).
60. G. I. Taylor, On making holes in a sheet of fluid, *J. Fluid Mech.*, **58**, 625 (1973).
61. D. H. Michael, Meniscus stability, *Ann. Rev. Fluid Mech.*, **13**, 189 (1981).
62. M. V. Berry, Cusped rainbows and incoherence effects in the rippling-mirror model for particle scattering from surfaces, *J. Phys.*, Ser. A, **8**, 566 (1975).
63. M. V. Berry, Waves and Thom's theorem, *Adv. Phys.*, **25**, 1 (1976).
64. J. F. Nye, Optical caustics in the near field from liquid drops, *Proc. Roy. Soc. Lond.*, Ser. A, **361**, 21 (1978).
65. M. V. Berry, J. F. Nye, and F. J. Wright, The elliptic umbilic diffraction catastrophe, *Phil. Trans. Roy. Soc.*, **291** (1382), 453 (1979).
66. J. F. Nye, Optical caustics from liquid drops under gravity: observations of the parabolic and symbolic umbilics, *Phil. Trans. Roy. Soc.*, Ser. A, **292** (1387), 25 (1979).
67. M. V. Berry, Focusing and twinkling: critical exponents from catastrophes in non-gaussian random short waves, *J. Phys.*, Ser. A, **10**, 2061 (1977).
68. A. S. Thorndike, C. R. Cooley, and J. F. Nye, The structure and evolution of flow fields and other vector fields, *J. Phys.*, Ser. A, **11**, 1455 (1978).
69. J. F. Nye and A. S. Thorndike, Events in evolving three-dimensional vector fields, *J. Phys.*, Ser. A, **13**, 1 (1980).
70. J. M. T. Thompson, J. K. Y. Tan, and K. C. Lim, On the topological classification of post-buckling phenomena, *J. Struct. Mech.*, **6**, 383 (1978).
71. G. Wassermann, Stability of Unfoldings in space and time, *Acta Mathematica*, **135**, 57 (1975).
72. G. Wassermann, *Stability of Unfoldings*, Springer Lecture Notes in Mathematics no. 393, Springer, Berlin, 1975.
73. G. Wassermann, (r,s)-stable unfoldings and catastrophe theory, *Structural Stability, The Theory of Catastrophes, and Applications in the Sciences* (Ed. P. Hilton), Lecture Notes in Mathematics No. 525, Springer, Berlin, 1976.

74. M. Golubitsky and D. Schaeffer, An analysis of imperfect bifurcation, *Annals, New York Academy of Sciences*, **316**, 127 (1979).
75. M. Golubitsky and D. Schaeffer, A theory for imperfect bifurcation via singularity theory, *Commun. Pure Appl. Math.*, **32**, 21 (1979).
76. M. Golubitsky and D. Schaeffer, Imperfect bifurcation in the presence of symmetry, *Commun. Math. Phys.*, **67**, 205 (1979).
77. J. M. T. Thompson and G. W. Hunt, A bifurcation theory for the instabilities of optimization and design, *Synthese*, **36**, 315 (1977).
78. G. W. Hunt, Imperfections and near-coincidence for semi-symmetric bifurcations, *Proc. Conf. on Bifurcation Theory and Applications in Scientific Disciplines, New York, October, 1977, in Annals, New York Acad. Sci.*, **316**, 572 (1979).
79. G. W. Hunt, Imperfection-sensitivity of semi-symmetric branching, *Proc. Roy. Soc. Lond.*, Ser. A, **357**, 193 (1977).
80. G. W. Hunt, N. A. Reay, and T. Yoshimura, Local diffeomorphisms in the bifurcational manifestations of the umbilic catastrophes, *Proc. Roy. Soc. Lond.*, Ser. A, **369**, 47 (1979).
81. G. W. Hunt, An algorithm for the nonlinear analysis of compound bifurcation, *Phil. Trans. Roy. Soc.*, Ser. A, **300** (1455), 443 (1981).
82. J. M. T. Thompson and P. A. Shorrock, Bifurcational instability of an atomic lattice, *J. Mech. Phys. Solids*, **23**, 21 (1975).
83. J. M. T. Thompson and P. A. Shorrock, Hyperbolic umbilic catastrophe in crystal fracture, Letter to *Nature*, **260**, 598 (1976).
84. M. V. Berry and M. R. Mackley, The six roll mill: unfolding an unstable persistently extensional flow, *Phil. Trans. Roy. Soc. Lond.*, Ser. A, **287**, 1 (1977).
85. M. R. Mackley, Flow singularities, polymer chain extension and hydrodynamic instabilities, *J. Non-Newtonian Fluid Mech.*, **4**, 111 (1978).
86. A. H. Chilver, Coupled modes of elastic buckling, *J. Mech. Phys. Solids*, **15**, 15 (1967).
87. J. M. T. Thompson and W. J. Supple, Erosion of optimum designs by compound branching phenomena, *J. Mech. Phys. Solids*, **21**, 135 (1973).
88 L. Bauer, H. Keller, and E. Reiss, Multiple eigenvalues lead to secondary bifurcation, *SIAM J. Appl. Math.*, **17**, 101 (1975).
89. J. P. Keener, Perturbed bifurcation theory at multiple eigenvalues, *Arch. Rational Mech. Anal.*, **56**, 348 (1974).
90. J. M. T. Thompson, The rotationally-symmetric branching behaviour of a complete spherical shell, *Proc. K. ned. Akad. Wet.*, Ser. B, **67**, 295 (1964).
91. W. T. Koiter, The non-linear buckling problem of a complete spherical shell under uniform external pressure, *Proc. K. ned. Akad. Wet.*, Ser. B, **72**, 40 (1969).
92. J. M. T. Thompson, The elastic instability of a complete spherical shell, *Aero. Quart.*, **13**, 189 (1962).
93. J. M. T. Thompson, The post-buckling of a spherical shell by computer analysis, in *World Conference on Shell Structures* (Eds. S. J. Medwadowski *et al.*), National Academy of Sciences, Washington, 1964.
94. E. C. Zeeman, The classification of elementary catastrophes of codimension less than or equal to five, *Structural Stability, The Theory of Catastrophes, and Applications in the Sciences*, (Ed. P. Hilton), Lecture Notes in Mathematics no. 525, Springer, Berlin, 1976.
95. G. V. Parkinson and N. P. H. Brooks, On the aeroelastic instability of bluff cylinders, *J. Appl. Mech.*, **28**, 252 (1961).
96. G. V. Parkinson and J. D. Smith, The square prism as an aeroelastic non-linear oscillator, *Quart. J. Mech. and Appl. Math.*, **17**, 225 (1964).
97. M. Novak, Galloping and vortex induced oscillations of structures, *Proc. third International Conf. on Wind Effects on Buildings and Structures*, Science Council of Japan, *Tokyo, 1971*.

98. R. D. Blevins, *Flow-induced Vibration*, Van Nostrand, New York, 1977.
99. G. Nicolis, Patterns of spatio-temporal organization in chemical and biochemical kinetics, *SIAM-AMS Proc.*, **8**, 33 (1974).
100. G. Nicolis and I. Prigogine, *Self-Organization in Non-Equilibrium Systems. From Dissipative Structures to Order through Fluctuations*, Wiley, New York, 1977.
101. A. T. Winfree, Rotating chemical reactions, *Sci. Am.*, **230**, 82 (June 1974).
102. D'Arcy W. Thompson, *On Growth and Form*, The University Press, Cambridge, 1971.
103. J. M. T. Thompson, An evolution game for a prey–predator ecology, *Bull., Inst. Maths and Its Applics*, **15**, 162 (1979).
104. C. H. Waddington, A catastrophe theory of evolution, *Annals, New York Academy of Sciences*, **231**, 32 (1974).
105. M. M. Dodson, Quantum evolution and the fold catastrophe, *Evolutionary Theory*, **1**, 107 (1975).
106. R. M. May, Bifurcations and dynamic complexity in ecological systems, *Annals, New York Academy of Sciences*, **316**, 517 (1979).
107. W. D. Iwan and R. D. Blevins, A model for vortex-induced oscillation of structures, *J. Appl. Mech.*, **41**, 581 (1974).
108. R. T. Hartlen and I. G. Currie, Lift oscillation model for vortex-induced vibration, *J. Eng. Mech. Div., Am. Soc. Civ. Engrs*, **96**, 577 (1970).
109. A. B. Poore and A. Al-Rawi, Some applicable Hopf bifurcation formulas and an application in wind engineering, *Annals, New York Academy of Sciences*, **316**, 590 (1979).
110. E. H. Dowell, Non-linear oscillations of a fluttering plate, *AIAA J.*, **4**, 1267 (1966).
111. E. H. Dowell, *Aeroelasticity of Plates and Shells*, Noordhoff, Leyden, 1975.
112. Y. C. Fung, *An Introduction to the Theory of Aeroelasticity*, Wiley, New York, 1955.
113. P. Holmes, Bifurcations to divergence and flutter in flow induced oscillations: a finite dimensional analysis, *J. Sound Vibr.*, **53**, 471 (1977).
114. T. B. Benjamin, Dynamics of a system of articulated pipes conveying fluid, I. Theory, II. Experiments, *Proc. Roy. Soc. Lond.,* Ser. A, **261**, 457–486 and 487–499 (1961).
115. T. S. Lunn, Flow-induced Instabilities of Fluid-conveying Pipes, Ph.D. Thesis, University College London, 1981.
116. J. Roorda and S. Nemat-Nasser, An energy method for stability analysis of non-linear, non-conservative systems, *AIAA J.*, **5**, 1262 (1967).
117. J. Rousselet and G. Herrmann, Flutter of articulated pipes at finite amplitude, *J. Appl. Mech.*, **44**, 154 (1977).
118. R. E. D. Bishop and I. Fawzy, Free and forced oscillation of a vertical tube containing a flowing fluid, *Phil. Trans. Roy. Soc. Lond.*, Ser. A, **284** (1316), 1 (1976).
119. M. P. Paidoussis and E. B. Deksnis, Articulated models of cantilevers conveying fluid: the study of a paradox, *J. Mech. Engng Sci.*, **12**, 288 (1970).
120. W. T. Koiter, On the instability of equilibrium in the absence of a minimum of the potential energy, *Proc. K. ned. Akad. Wet.*, Ser. B, **68**, 107 (1965).
121. R. C. T. Rainey, The dynamics of tethered platforms, *Trans. Roy. Inst. Naval Architects*, **120**, 59 (1978).
122. R. C. T. Rainey, Parasitic motions of offshore structures, *Trans. Roy. Inst. Naval Architects*, in press. (Paper for written discussion, W3, 1980.)
123. J. R. Richardson, *Mathieu Instabilities and Response of Compliant Offshore Structures*, National Marine Institute, NMIR 49 (OT-R-7913), February 1979.
124. P. J. Holmes and D. A. Rand, The bifurcations of Duffing's equations: an application of catastrophe theory, *J. Sound Vib.*, **44**, 237 (1976).
125. J. E. Marsden, Qualitative methods in bifurcation theory, *Bull. Am. Math. Soc.*, **84**, 1125 (1978).
126. P. J. Holmes and J. Marsden, Bifurcations to divergence and flutter in flow-induced oscillations: an infinite dimensional analysis, *Automatica*, **14**, 367 (1978).

127. P. Holmes and J. E. Marsden, Qualitative techniques for bifurcation analysis of complex systems, *Annals, New York Academy of Sciences*, **316**, 608 (1979).
128. P. J. Holmes and Y. K. Lin, Deterministic stability analysis of a wind loaded structure, *J. Appl. Mech.*, **45**, 165 (1978).
129. Y. K. Lin and P. J. Holmes, Stochastic analysis of a wind loaded structure, *J. Eng. Mech. Div. Am. Soc. Civ. Engs.*, **104**, 421 (1978).
130. D. R. J. Chillingworth and P. J. Holmes, Dynamical systems and models for reversals of the earth's magnetic field, *J. Math. Geology*, **12**, 41 (1980).
131. P. J. Holmes and D. A. Rand, Bifurcations of the forced Van der Pol oscillator, *Quart. Appl. Math.*, **35**, 495 (1978).
132. P. Holmes and D. Rand, Identification of vibrating systems by generic modelling, *Proc. Int. Symp. on Shipboard Acoustics Delft* (Ed. J. H. Janssen), Elsevier, 1977.
133. P. J. Holmes and J. E. Marsden, Bifurcations of dynamical systems and nonlinear oscillators in engineering systems, *Nonlinear partial differential equations and applications*, Lecture Notes in Mathematics no. 648, p. 163 Springer, Berlin, 1978.
134. P. J. Holmes (Ed.), *New Approaches to Nonlinear Problems in Dynamics*, Proc. Conf Pacific Grove, 1979 (SIAM, Philadelphia, 1980).
135. O. Gurel and O. E. Rossler (Eds), Bifurcation theory and applications in scientific disciplines, *Annals, New York Academy of Sciences*, **316** (1979).
136. S. N. Chow, J. K. Hale, and J. Mallet-Paret, Applications of generic bifurcation, I and II, *Arch. Rat. Mech. Anal.*, **59**, 159 (1975) and **62**, 209 (1976).
137. J. K. Hale, *Generic Bifurcation with Applications, Heriot-Watt Lecture*, Vol. I, Pitman, London, 1977.
138. M. Potier-Ferry, Bifurcation et stabilité pour des systèmes dérivant d'un potentiel, *Journal de Mécanique*, **17**, 38 (1978).
139. M. Potier-Ferry, Perturbed bifurcation theory, *J. Diff. Equations*, **33**, 112 (1979).
140. M. S. El Naschie and S. Al Athel, On the morphology of controlled systems, *Proc. IUTAM Symposium on Structural Control*, North Holland Publishing Co., Amsterdam, 1980.
141. W. Guttinger and H. Eikemeier (Eds), *Structural Stability in Physics*, Springer-Verlag, Berlin, 1979.
142. D. R. J. Chillingworth, *Differential Topology with a View to Applications*, Research Notes in Mathematics no. 9, Pitman, London, 1976.
143. G. Herrmann and I. Jong, On the destabilizing effect of damping in non-conservative elastic systems, *J. Appl. Mech.*, **32**, 592 (1965).
144. S. Nemat-Nasser, S. N. Prasad, and G. Herrmann, Destabilizing effect of velocity-dependent forces in nonconservative continuous systems, *AIAA J.*, **4**, 1276 (1966).
145. E. N. Lorenz, Deterministic non-periodic flow, *J. Atmos. Sciences*, **20**, 130 (1963).
146. D. Ruelle and F. Takens, On the nature of turbulence, *Comm. Math. Phys.*, **20**, 167 (1971).
147. O. E. Rossler, Continuous chaos—four prototype equations, *Annals, New York Academy of Sciences*, **316**, 376 (1979).
148. O. E. Rossler, Chaos, in *Structural Stability in Physics* (Eds W. Guttinger and H. Eikemeier), p. 290, Springer, Berlin, 1978.
149. P. J. Holmes, Strange phenomena in dynamical systems and their physical implications, *Appl. Math. Modelling*, **1**, 362 (1977).
150. F. Moon and P. J. Holmes, A magneto-elastic strange attractor, *J. Sound Vibr.*, **65**, 275 (1979).
151. P. J. Holmes, A nonlinear oscillator with a strange attractor, *Phil. Trans. Roy. Soc. Lond.*, Ser. A, **292**, 419 (1979).
152. P. J. Holmes, Averaging and chaotic motions in forced oscillations, *SIAM J. Appl. Math.*, **38**, 65 (1980).
153. P. J. Holmes, Global bifurcations and chaos in the forced oscillations of buckled structures, *Proc. IEEE Conf. on Decision and Control, San Diego, 10–12 January 1979.* Inst. Elec. Electronics Engrs., New York.

154. J. E. Marsden and P. J. Holmes, A horseshoe in the dynamics of a forced beam, *Proc. Int. Conf. on Nonlinear Dynamics*, New York, 1979 (New York Academy of Sciences, 1980).

155. P. J. Holmes and F. Moon, Addendum: a magnetoelastic strange attractor, *J. Sound and Vibr.*, **69**, 339 (1980).

156. P. J. Holmes, Periodic, nonperiodic and irregular motions in a Hamiltonian system, to appear.

157. I. Gumowski and C. Mira, *Dynamique Chaotique*, Cepadues Editions, Toulouse, 1980.

158. M. Hénon. A two-dimensional mapping with a strange attractor, *Comm. Math. Phys.*, **50**, 69 (1976).

159. C. S. Hsu and H. C. Yee, Behaviour of dynamical systems governed by a simple nonlinear difference equation, *J. Appl. Mech.*, **42**, 870 (1975).

160. C. S. Hsu and W. H. Cheng, Steady-state response of a nonlinear system under impulsive periodic parametric excitation, *J. Sound and Vibr.*, **50**, 95 (1977).

161. C. S. Hsu, H. C. Yee, and W. H. Cheng, Determination of global regions of asymptotic stability for difference dynamical systems, *J. Appl. Mech.*, **44**, 147 (1977).

162. C. Lanczos, *The Variational Principles of Mechanics*, 2nd ed., The University Press, Toronto, 1962.

163. D. Chillingworth, The catastrophe of a buckling beam, in *Dynamical Systems, Warwick 1974* (Ed. A. Manning), Lecture Notes in Mathematics no. 468, Springer, Berlin, 1975.

164. E. C. Zeeman, Euler buckling, in *Structural Stability, The Theory of Catastrophes, and Applications in the Sciences* (Ed. P. Hilton), Lecture Notes in Mathematics no. 525, Springer, Berlin, 1976.

165. M. S. El Naschie, Exact asymptotic solution for the initial post-buckling of a strut on a linear elastic foundation, *ZAMM.*, **54**, 677 (1971).

166. M. S. El Naschie, Zum knickmechanismus des idealen kreisringes, *Der Stahlbau*, **1**, 23 (1976).

167. J. M. T. Thompson, Discrete branching points in the general theory of elastic stability, *J. Mech. Phys. Solids*, **13**, 295 (1965).

168. J. M. T. Thompson, The branching analysis of perfect and imperfect discrete structural systems, *J. Mech. Phys. Solids*, **17**, 1 (1969).

169. Y. Hangai and S. Kawamata, Analysis of geometrically nonlinear and stability problems by static perturbation method, *Rept. Inst. Indust. Sci. Univ. Tokyo*, **22**, No. 5, 1973.

170. S. J. Britvec and A. H. Chilver, Elastic buckling of rigidly-jointed braced frames, *J. Eng. Mech. Div., Am. Soc. Civ. Engrs.*, **89**, 217 (1963).

171. S. J. Britvec, *The Stability of Elastic Systems*, Pergamon, New York, 1973.

172. J. M. T. Thompson and Z. Gaspar, A buckling model for the set of umbilic catastrophes, *Math. Proc. Camb. Phil. Soc.*, **82**, 497 (1977).

173. E. C. Zeeman, The umbilic bracelet and the double-cusp catastrophe, in *Structural Stability, The Theory of Catastrophes, and Applications in the Sciences* (Ed. P. Hilton), Lecture Notes in Mathematics no. 525, Springer, Berlin, 1976.

174. D. Ho, Higher order approximations in the calculation of elastic buckling loads of imperfect systems, *Int. J. Non-linear Mech.*, **6**, 649 (1971).

175. D. Ho, The influence of imperfections on systems with coincident buckling loads, *Int. J. Non-linear Mech.*, **7**, 311 (1972).

176. D. Ho, Buckling load of nonlinear systems with multiple eigenvalues, *Int. J. Solids Struct.*, **10**, 1315 (1974).

177. J. M. T. Thompson, Bifurcational aspects of catastrophe theory, *Proc. Conf. on Bifurcation Theory and Applications in Scientific Disciplines, New York, October 1977*, in *Annals, New York Academy of Science*, **316**, 553 (1979).

178. K. Huseyin, The multiple-parameter stability theory and its relation to catastrophe theory, in *Problem Analysis in Science and Engineering*, Academic Press, New York, 1977.
179. J. S. Hansen, Some two-mode buckling problems and their relation to catastrophe theory, *AIAA J.*, **15**, 1638 (1977).
180. D. Hui and J. S. Hansen, The swallowtail and butterfly cuspoids and their application in the initial post-buckling of single-mode structural systems, *Q. Appl. Math.*, **38**, 17 (1980).
181. P. Samuels, The relationship between postbuckling behaviour at coincident branching points and the geometry of an umbilic point of the energy surface, *J. Struct. Mech.*, **7**, 297 (1979).
182. P. Samuels, Bifurcation and limit point instability of dual eigenvalue third order systems, *Int. J. Solids Struct.*, **16**, 743 (1980).
183. D. Schaeffer and M. Golubitsky, Boundary conditions and mode jumping in the buckling of a rectangular plate, *Commun. Math. Phys.*, **69**, 209 (1979).
184. D. R. J. Chillingworth, Universal bifurcation problems in mechanics of solids, in *The Rodney Hill Sixtieth Anniversary Volume* (Ed. H. G. Hopkins and M. J. Sewell), Pergamon, Oxford, 1981.
185. R. J. Knops and E. W. Wilkes, Theory of elastic stability, in *Handbuch der Physik* (Ed. S. Flugge) Vol. VIa/3, Springer-Verlag, Berlin, 1973.
186. T. von Kármán and H. S. Tsien, The buckling of spherical shells by external pressure, *J. aeronaut. Sci.*, **7**, 43 (1939).
187. A. Van der Neut, The interaction of local buckling and column failure of thin-walled compression members, *Proc. Twelfth International Cong. Appl. Mech., Stanford, 1968*, Springer-Verlag, Berlin, 1968.
188. J. M. T. Thompson and G. M. Lewis, On the optimum design of thin-walled compression members, *J. Mech. Phys. Solids*, **20**, 101 (1972).
189. R. B. Gilbert and C. R. Calladine, Interaction between the effects of local and overall imperfections on the buckling of elastic columns, *J. Mech. Phys. Solids*, **22**, 519 (1974).
190. S. E. Svensson and J. G. A. Croll, Interaction between local and overall buckling, *Int. J. Mech. Sci.*, **17**, 307 (1975).
191. A. C. Walker, Interactive buckling of structural components, *Sci. Prog. Oxf.*, **62**, 579 (1975).
192. J. M. T. Thompson, J. D. Tulk, and A. C. Walker, An experimental study of imperfection-sensitivity in the interactive buckling of stiffened plates, in *Buckling of Structures* (Ed. B. Budiansky), Springer-Verlag, Berlin, 1976.
193. R. Maquoi and Ch. Massonnet, Interaction between local plate buckling and overall buckling in thin-walled compression members—theories and experiments, in *Buckling of Structures* (Ed. B. Budiansky), Springer-Verlag, Berlin, 1976.
194. E. Byskov and J. W. Hutchinson, Mode interaction in axially stiffened cylindrical shells, *AIAA J.*, **15**, 941 (1977).
195. W. T. Koiter and A. Van der Neut, Interaction between local and overall buckling of stiffened compression panels, in *Thin-Walled Structures* (Eds J. Rhodes and A. C. Walker), Granada, London, 1980.
196. J. G. A. Croll, Towards simple estimates of shell buckling loads, *Der Stahlbau*, **8**, 243 and **9**, 283 (1975).
197. R. C. Batista and J. G. A. Croll, A design approach for axially compressed un-stiffened cylinders, in *Stability Problems in Engineering Structures and Components* (Eds T. H. Richards and P. Stanley), Applied Sci. Publishers, London, 1979.
198. C. P. Ellinas and J. G. A. Croll, The basis of a design approach for stiffened plates, in *Stability Problems in Engineering Structures and Components* (Eds T. H. Richards and P. Stanley), Applied Sci. Publishers, London, 1979.

208

199. R. C. Batista and J. G. A. Croll, A design approach for unstiffened cylindrical shells under external pressure, in *Thin-Walled Structures* (Eds J. Rhodes and A. C. Walker), Granada, London, 1980.
200. J. M. T. Thompson and G. W. Hunt, A theory for the numerical analysis of compound branching, *Z. angew. Math. Phys.*, **22**, 1001 (1971).
201. E. Riks, *The Incremental Solution of some Basic Problems in Elastic Stability*, National Aerospace Laboratory, NLR, TR 74005 U, 1973.
202. J. F. Besseling, Post-buckling and non-linear analysis by the finite element method as a supplement to a linear analysis, *ZAMM.*, **55**, 3 (1975).
203. H. Fujii and M. Yamaguti, *Structure of Singularities and its Numerical Realization in Nonlinear Elasticity*, Inst. Computer Sic, Kyoto Sangyo University, Research Report, KSU/ICS 79–09, September 1979.
204. A. Rosen and L. A. Schmit, Design oriented analysis of imperfect truss structures, Part I: Accurate analysis, Part II: Approximate analysis, *Int. J. Num. Meth. Engng.*, **14**, 1309 (1979) and **15**, 483 (1980).
205. N. J. Hoff, Theory and experiment in the creep buckling of plates and shells, in *Buckling of Structures* (Ed. B. Budiansky), Springer-Verlag, Berlin, 1976.
206. B. Hayman, Aspects of creep buckling, I. The influence of post-buckling characteristics, II. The effects of small deflexion approximations on predicted behaviour, *Proc. Roy. Soc. Lond.*, Ser. A, **364**, 393–414 and 415–433 (1978).
207. J. W. Hutchinson and W. T. Koiter, Post-buckling theory, *Appl. Mech. Rev.*, **23**, 1353 (1970).
208. W. T. Koiter, Current trends in the theory of buckling, in *Buckling of Structures* (Ed. B. Budiansky), Springer-Verlag, Berlin, 1976.
209. B. Budiansky and J. W. Hutchinson, Buckling: progress and challenge, in *Trends in Solid Mechanics* (Eds J. F. Besseling and A. M. A. Van der Heijden), the University Press, Delft, 1979 (Proceedings of the Symposium dedicated to the sixty-fifth birthday of W. T. Koiter).
210. W. T. Koiter, Forty years in retrospect, the bitter and the sweet, in *Trends in Solid Mechanics* (Eds J. F. Besseling and A. M. A. Van der Heijden), The University Press, Delft, 1979 (Proceedings of the Symposium dedicated to the sixty-fifth birthday of W. T. Koiter).
211. V. Gioncu and M. Ivan, *Buckling of Shell Structures* (in Rumanian), Editura Academiei Republicii Socialiste Romania, Bucuresti, 1978.
212. V. Gioncu, *Thin Reinforced Concrete Shells*, Wiley, Chichester, 1979.
213. J. Rhodes and A. C. Walker (Eds), *Thin-Walled Structures*, Granada, London, 1980.
214. H. G. Allen and P. S. Bulson, *Background to Buckling*, McGraw-Hill, London, 1980.
215. A. Einstein, Zur elektrodynamik bewegter korper, *Ann. Phys.*, **17**, 891 (1905). English translation, The electrodynamics of moving bodies, in *The Principle of Relativity: Original Papers* (A. Einstein and H. Minkowski), The University Press, Calcutta, 1920.
216. A. Einstein, Die grundlagen der allgemeinen relativitaetstheorie, *Ann. Phys.*, **49**, 769 (1916). English translation, The foundation of the generalised theory of relativity, in *The Principle of Relativity: Original Papers* (A. Einstein and H. Minkowski), The University Press, Calcutta, 1920.
217. A. M. Liapunov, Sur les figures d'equilibre peu différentes des ellipsoides d'une masse liquide homogène douée d'un mouvement de rotation, *Zap. Akad. Nauk. St. Petersburg*, **1**, 1 (1906).
218. B. K. Harrison, K. S. Thorne, M. Wakano, and J. A. Wheeler, *Gravitation Theory and Gravitational Collapse*, The University Press, Chicago, 1965.
219. L. Landau, On the theory of stars (in English), *Physikalische Zeitschrift der Sowjetunion*, **1**, 285 (1932).

220. S. Chandrasekhar, The highly collapsed configurations of a stellar mass, *Mon. Not. Roy. Astr. Soc.*, **95**, 207 (1935).
221. S. Chandrasekhar, *An Introduction to the Study of Stellar Structure*, Dover, New York, 1958.
222. J. R. Oppenheimer and G. M. Volkoff, On massive neutron cores, *Phys. Rev.*, **55**, 374 (1939).
223. C. W. Misner and H. S. Zapolsky, High-density behaviour and dynamical stability of neutron star models, *Phys. Rev. Letters*, **12**, 635 (1964).
224. J. A. Wheeler, *Geometrodynamics*, Academic Press, New York, 1962.
225. J. A. Wheeler, Geometrodynamics and the issue of the final state, in *Relativity Theory, Groups and Topology* (Ed. B. De Witt), Gordon and Breach, London, 1964.
226. S. Weinberg, *Gravitation and Cosmology: Principles and Applications of the General Theory of Relativity*, Wiley, New York, 1972.
227. S. W. Hawking and G. F. R. Ellis, *The Large Scale Structure of Space-Time*, University Press, Cambridge, 1973.
228. C. W. Misner, K. S. Thorne, and J. A. Wheeler, *Gravitation*, Freeman, San Francisco, 1973.
229. M. Rees, R. Ruffini, and J. A. Wheeler, *Black Holes, Gravitational Waves and Cosmology: An Introduction to Current Research*, Topics in Astrophysics and Space Physics no. 10, Gordon and Breach, New York, 1974.
230. D. Lynden-Bell and R. Wood, The gravo-thermal catastrophe in isothermal spheres and the onset of red-giant structure for stellar systems, *Mon. Not. Roy. Astr. Soc.*, **138**, 495 (1968).
231. S. J. Aarseth, Dynamical evolution of clusters of galaxies, *Mon. Not. Roy. Astr. Soc.*, **126**, 223 (1963).
232. V. A. Antonov, The most probable phase distribution in spherical stellar systems and the conditions of its existence, *Vestnik Leningradskogo Universiteta, Seriya Matematiki Mekhaniki i Astronomii*, **17**, No. 7, 135 (1962).
233. P. Glansdorff and I. Prigogine, *Thermodynamic Theory of Structure, Stability and Fluctuations*, Wiley, London, 1971.
234. J. Katz, On the number of unstable modes of an equilibrium, *Mon. Not. Roy. Astr. Soc.*, **183**, 765 (1978).
235. J. Katz, On the number of unstable modes of an equilibrium II, *Mon. Not. Roy. Astr. Soc.*, **189**, 817 (1979).
236. J. Katz, Stability limits for isothermal cores in globular clusters, *Mon. Not. Roy. Astr. Soc.*, **190**, 497 (1980).
237. G. H. Darwin, *Scientific Papers*, Vol. III, *Figures of Equilibrium of Rotating Liquid and Geophysical Investigations*, Cambridge University Press, Cambridge, 1910.
238. J. H. Jeans, *Problems of Cosmogony and Stellar Dynamics*, Cambridge University Press, Cambridge, 1919.
239. J. H. Jeans, *Astronomy and Cosmogony*, Cambridge University Press, Cambridge, 1928.
240. R. A. Lyttleton, *The Stability of Rotating Liquid Masses*, Cambridge University Press, Cambridge, 1953.
241. P. Ledoux, Stellar stability, in *Handbuch der Physik*, (Ed. S. Flugge), Vol. LI, Springer-Verlag, Berlin, 1958.
242. S. Chandrasekhar, *Ellipsoidal Figures of Equilibrium*, Yale University Press, New Haven, 1969.
243. J. W. Hutchinson and V. Tvergaard, Shear band formation in plane strain, *Int. J. Solids Structures*, **17**, 451 (1981).
244. N. H. Macmillan and A. Kelly, The mechanical properties of perfect crystals, *Proc. Roy. Soc.*, Ser. A, **330**, 291 (1972).
245. M. Born and K. Huang, *Dynamical Theory of Crystal Lattices*, University Press, Oxford, 1954.

246. P. Dean, Atomic vibrations in solids, *J. Inst. Maths. Applics*, **3**, 98 (1967).
247. R. Hill, On the elasticity and stability of perfect crystals at finite strain, *Math. Proc. Camb. Phil. Soc.*, **77**, 225 (1975).
248. R. Hill and F. Milstein, Principles of stability analysis of ideal crystals, *Physical Review*, Ser. B, **15**, 3087 (1977).
249. C. A. Coulson, The role of mathematics in chemistry, *IMA Bull.*, **9**, 206 (1973).
250. R. D. Mindlin, Elasticity, piezoelectricity and crystal lattice dynamics, *J. Elasticity*, **2**, 217 (1972).
251. R. Hill and J. W. Hutchinson, Bifurcation phenomena in the plane tension test, *J. Mech. Phys. Solids*, **23**, 239 (1975).
252. G. Nicolis, Irreversible thermodynamics, *Reports on Progress in Physics*, **42**, 225 (1979).
253. J. F. G. Auchmuty and G. Nicolis, Bifurcation analysis of nonlinear reaction-diffusion equations, I, Evolution equations and the steady state solutions, *Bull. Math. Biol.*, **37**, 323 (1975).
254. T. Erneux and M. Herschkowitz-Kaufman, Dissipative structures in two dimensions, *Biophys. Chem.*, **3**, 345 (1975).
255. B. C. Goodwin, The analysis of rythmic behaviour in organisms: a phenomenological approach, *IMA Bull.*, **12**, 2 (1976).
256. T. Erneux and J. Hiernaux, Transition from polar to duplicate patterns, *J. Math. Biology*, **9**, 193 (1980).
257. L. Wolpert, Positional information and the spatial pattern of cellular differentiation, *J. Theor. Biology*, **25**, 1 (1969).
258. T. Erneux, J. Hiernaux, and G. Nicolis, Turing's theory of morphogenesis, *Bull. Math. Biology*, **40**, 771 (1978).
259. A. M. Turing, The chemical basis of morphogenesis, *Phil. Trans. Roy. Soc. Lond.*, Ser. B, **237**, 37 (1952).
260. G. Nicolis and M. Malek-Mansour, Non-equilibrium phase transitions and chemical reactions, *Progress of Theoretical Physics*, Supplement No. 64, 249 (1978).
261. R. M. May (Ed.), *Theoretical Ecology: Principles and Applications*, Blackwell, Oxford, 1976.
262. J. T. Tanner, The stability and the intrinsic growth rates of prey and predator populations, *Ecology*, **56**, 855 (1975).
263. R. M. May, Thresholds and breaking points in ecosystems with a multiplicity of stable states, *Nature*, **269**, 471 (1977).
264. O. Reynolds, *Papers on Mechanical and Physical Subjects* (reprinted from various Transactions and Journals, in three Volumes), Cambridge University Press, Cambridge, 1900–1903.
265. M. J. Lighthill, Turbulence, Chapter 2 in *Osborne Reynolds and Engineering Science Today* (Eds D. M. McDowell and J. D. Jackson), Manchester University Press, Manchester, 1970.
266. D. D. Joseph, *Stability of Fluid Motions*, Vols I and II, Springer Tracts in Natural Philosophy Volumes 27 and 28, Springer Verlag, Berlin, 1976.
267. C. C. Lin, *The theory of Hydrodynamic Stability*, University Press, Cambridge, 1967.
268. G. I. Taylor, Stability of a viscous liquid contained between two rotating cylinders, *Phil. Trans. Roy. Soc. Lond.*, Ser. A, **223**, 289–343 (1923).
269. S. Chandrasekhar, *Hydrodynamic And Hydromagnetic Stability*, The University Press, Oxford, 1968.
270. P. Bernard and T. Ratiu (Eds), *Turbulence Seminar, Berkeley 1976/7*, Springer Lecture Notes in Mathematics, no. 615, Springer, Berlin, 1977.
271. J. L. Kaplan and J. A. Yorke, The onset of chaos in a fluid flow model of Lorenz, *Annals, New York Academy of Sciences*, **316**, 400 (1979).
272. J. M. T. Thompson and R. J. Thompson, Numerical experiments with a strange attractor, *Bull. Inst. Maths and its Applics*, **16**, 150 (1980).

273. D. Ruelle, Strange attractors, *The Mathematical Intelligencer*, **2**, 129 (1980).
274. J. M. T. Thompson, Static and dynamic instabilities in the physical sciences: an inaugural lecture, *J. Eng. Sci., Univ. Riyadh*, **6**, 71 (1980).
275. N. Krylov and N. N. Bogoliubov, *Introduction to Non-linear Mechanics*, translated by S. Lefschetz, Annals of Maths Studies no. 11, University Press, Princeton, 1947.
276. A. A. Andronov and C. E. Chaikin, *Theory of Oscillations*, University Press, Princeton, 1949.
277. A. A. Andronov, A. A. Vitt, and S. E. Khaiken, *Theory of Oscillators*, English ed., Pergamon Press, Oxford, 1966.
278. A. A. Andronov, E. A. Leontovich, I. I. Gordon, and A. G. Maier, *Theory of Bifurcations of Dynamical Systems on a Plane* and *Qualitative Theory of Second-order Dynamical Systems*, Wiley, New York, 1971 and 1973.
279. J. La Salle and S. Lefschetz, *Stability by Liapunov's Direct Method with Applications*, Academic Press, New York, 1961.
280. V. V. Bolotin, *Nonconservative Problems of the Theory of Elastic Stability*, Pergamon Press, Oxford, 1963.
281. V. V. Bolotin, *The Dynamic Stability of Elastic Systems*, Holden-Day, San Francisco, 1964.
282. C. Hayashi, *Nonlinear Oscillations in Physical Systems*, McGraw-Hill, New York, 1964.
283. L. A. Pars, *A Treatise on Analytical Dynamics*, Heinemann, London, 1965.
284. R. M. Rosenberg, On nonlinear vibration of systems with many degrees of freedom, *Adv. Appl. Mech.*, **9**, 159 (1966).
285. A. Blaquiere, *Nonlinear System Analysis*, Academic Press, New York, 1966.
286. D. W. Jordan and P. Smith, *Nonlinear Ordinary Differential Equations*, Clarendon Press, Oxford, 1977.
287. A. B. Pippard, *The Physics of Vibration*, Vol. I, The University Press, Cambridge, 1978.
288. A. H. Nayfeh and D. T. Mook, *Nonlinear Oscillations*, Wiley, New York, 1979.
289. E. Mettler, Dynamic buckling, Chapter 62 in *Handbook of Engineering Mechanics* (Ed. S. Flugge), McGraw-Hill, New York, 1962.
290. J. W. Hutchinson and B. Budiansky, Dynamic buckling estimates, *AIAA J.*, **4**, 525 (1966).
291. J. M. T. Thompson, Dynamic buckling under step loading, in *Dynamic Stability of Structures* (Ed. G. Herrmann), Pergamon Press, Oxford, 1966.
292. R. H. Plaut, Postbuckling analysis of nonconservative elastic systems, *J. Struct. Mech.*, **4**, 395 (1976).
293. R. H. Plaut, Branching analysis at coincident buckling loads of nonconservative elastic systems, *J. Appl. Mech.*, **44**, 317 (1977).
294. K. Huseyin, On the stability of equilibrium paths associated with autonomous systems, *J. Appl. Mech.*, Paper no. 81-APM-9. In press.
295. K. Huseyin and V. Mandadi, On the instability of multiple-parameter systems, Sectional Lecture, *Proc. Fifteenth International Cong. Theoretical and Appl. Mech., Toronto, 1980*, North-Holland Publishing Company, 1980.
296. I. Stewart, Applications of catastrophe theory to the physical sciences, *Physica* (D: Nonlinear Phenomena), **2**, 245 (1981).
297. D. Thornton, A general review of future problems and their solution, *Proc. Second International Conf. on Behaviour of Off-Shore Structures, BOSS'79, Imperial College, London, August, 1979*, BHRA Fluid Engineering, Cranfield, 1979.
298. A. Pugsley, *The Safety of Structures*, Arnold, London, 1966.
299. R. E. D. Bishop, *Vibration*, Cambridge University Press, Cambridge, 1979.
300. M. Novak, Aeroelastic galloping of prismatic bodies, *J. Eng. Mech. Div. Am. Soc. Civ. Engs*, **95**, 115 (1969).
301. F. Takens, Singularities of vector fields, *Publ. IHES*, **43**, 47 (1974).

212

302. F. Takens, *Forced oscillations and Bifurcations*, Communication No. 3, Mathematical Institute of Rijksuniversiteit, Utrecht, 1974.
303. G. T. S. Done, *The Effect of Linear Damping on Flutter Speed*, ARC Reports and Memoranda No. 3396, March 1963.
304. G. T. S. Done, *The Flutter and Stability of Undamped Systems*, ARC Reports and Memoranda No. 3553, November 1966.
305. J. M. T. Thompson and T. S. Lunn, Static elastica formulations of a pipe conveying fluid, *J. Sound and Vibr.*, **77**, 127 (1981).
306. J. M. T. Thompson, On the simulation of a gravitational field by a centrifugal field, *Int. J. Mech. Sci.*, **13**, 979 (1971).
307. M. P. Paidoussis and N. T. Issid, Dynamic stability of pipes conveying fluid, *J. Sound and Vibr.*, **33**, 267 (1974).
308. J. M. T. Thompson and T. S. Lunn, The non-linear dynamics of a simply-supported pipe conveying fluid, to appear.
309. F. Niordson, Vibrations of a cylindrical tube containing flowing fluid, *Acta Polytechnica, Mechanical Engineering Series*, **3**, No. 2, 1954.
310. A. L. Thurman and C. D. Mote, Nonlinear oscillation of a cylinder containing a flowing fluid, *J. Engng for Industry*, **91**, 1147 (1969).
311. K. Huseyin and R. H. Plaut, Transverse vibrations and stability of systems with gyroscopic forces, *J. Struct. Mech.*, **3**, 163 (1975).
312. R. H. Plaut and K. Huseyin, Instability of fluid-conveying pipes under axial load, *J. Appl. Mech.*, **42**, 889 (1975).
313. P. J. Holmes, Pipes supported at both ends cannot flutter, to appear.
314. D. S. Weaver, On the non-conservative nature of gyroscopic conservative systems, *J. Sound and Vibr.*, **36**, 435 (1974).
315. M. P. Paidoussis, Flutter of conservative systems of pipes conveying incompressible fluid, *J. Mech. Engng. Sci.*, **17**, 19 (1975).
316. G. Herrmann and R. W. Bungay, On the stability of elastic systems subjected to non-conservative forces, *J. Appl. Mech.*, **31**, 435 (1964).
317. G. Herrmann and S. Nemat-Nasser, Energy considerations in the analysis of stability of nonconservative structural systems, in *Dynamic Stability of Structures* (Ed. G. Herrmann), Pergamon Press, Oxford, 1966.
318. R. W. Gregory and M. P. Paidoussis, Unstable oscillation of tubular cantilevers conveying fluid, I. Theory, II. Experiments, *Proc. Roy. Soc. Lond.*, Ser. A, **293**, 512–527 and 528–542 (1966).
319. M. P. Paidoussis, Dynamics of tabular cantilevers conveying fluid, *J. Mech. Engng. Sci.*, **12**, 85 (1970).
320. I. W. Burgess and M. Levinson, The post-flutter oscillations of discrete symmetric structural systems with circulatory loading, *Int. J. Mech. Sci.*, **14**, 471 (1972).
321. M. P. Bohn and G. Herrmann, Instabilities of a spatial system of articulated pipes conveying fluid, *J. Fluids Engng.*, **96**, 289 (1974).
322. H. H. E. Leipholz, On the application of the energy method to the stability problem of nonconservative autonomous and nonautonomous systems, *Acta Mechanica*, **28**, 113 (1977).
323. B. Van der Pol, Forced oscillations in a circuit with nonlinear resistance, *Phil Mag.*, **3**, 65 (1927).
324. B. Van der Pol, The nonlinear theory of electric oscillations, *Proc. Inst. Radio Engrs.*, **22**, 1051 (1934).
325. J. M. T. Thompson and T. S. Lunn, Resonance-sensitivity in dynamic Hopf bifurcations under fluid loading, *Appl. Math. Modelling*, **5**, 143 (1981).
326. N. G. Chetayev, *The Stability of Motion* (translated from the Russian by M. Nadler), Pergamon Press, New York, 1961.
327. T. R. Kane and D. A. Levinson, Stability, instability, and terminal attitude motion of a spinning, dissipative spacecraft, *AIAA J.*, **14**, 39 (1976).

328. T. R. Kane and P. M. Barba, Effects of energy dissipation on a spinning satellite, *AIAA J.*, **4**, 1391 (1966).
329. T. R. Robe and T. R. Kane, Dynamics of an elastic satellite, I, II and III, *Int. J. Solids Structures*, **3**, 333–352, 691–703, and 1031–1051 (1967).
330. D. R. Teixeira and T. R. Kane, Spin stability of torque-free systems, *AIAA J.*, **11**, 862 (1973).
331. R. Pringle Jr, On the stability of a body with connected moving parts, *AIAA J.*, **4**, 1395 (1966).
332. T. R. Kane and D. R. Teixeira, Instability of rotation about a centroidal axis of maximum moment of inertia, *AIAA J.*, **10**, 1356 (1972).
333. C. E.N. Mazzilli, *A Class of Non-linear Vibrations and their Stability*, Ph.D. Thesis, University College London, 1981.
334. H. R. Wilson and J. D. Cowan, Excitatory and inhibitory interactions in localized populations of model neurons, *Biophysical J.*, **12**, 1 (1972).
335. H. R. Wilson, Mathematical models of neural tissue, in *Cooperative Effects, Progress in Synergetics* (Ed. H. Haken), North-Holland Publishing Co, Amsterdam, 1974.
336. P. A. Anninos, The usefulness of artificial neural nets as models for the normal and abnormal functioning of the mammalian CNS, *Progress in Neurobiology*, **4**, 57 (1975).
337. S. Amari, A mathematical approach to neural systems, in *Systems Neuroscience* (Ed. J. Metzler), Academic Press, New York, 1977.

Index

(See also the Index of Bifurcations that follows.)

Index of Bifurcations and Catastrophes

I. Non-linear static instabilities

(*see* Figure 10), xvi, 6, 7, 10, 57, 59, 136

CATASTROPHES	BIFURCATIONS	Alphabetical examples
FOLD (Saddle-node)	LIMIT POINT	Arch, deep (after bifurcation), 56 Arch, shallow, 12, 55 Atomic lattice, 88, 98 Forced oscillator, 182 General, 11, 32 Neurology, 193, 195 Spherical dome, 12 Stars, relativistic, 64, 66 Stellar clusters, 72, 74 Strut, complementary path, 35
	ASYMMETRIC POINT (Trans-critical)	Charged liquid drop, 12 Developmental biology, 111 Frames, structural, 57 General, 6, 12, 13, 57, 58, 136 Reaction-diffusion, 109 Turbulence in Couette flow, 140
CUSP	GENERAL	Caustics, 17, 18, 19, 20 Developmental biology, 111 General, 14, 16 Psychology of rage and fear, 17
	STABLE-SYMMETRIC (Super-critical)	Cantilevered link model, 31 General, 6, 14, 58, 136 Pipe divergence, simple supports, 29, 166 Plate buckling, 57 Reaction-diffusion, 109 Rotating liquid planets, 79 Strut, 15, 35, 37, 38, 43, 45, 57 Turbulence in Couette flow, 16, 140
	UNSTABLE-SYMMETRIC (Sub-critical)	Arch, deep, 14, 56 Atomic lattice, 81, 98 General, 6, 14, 59, 136 Rotating liquid planets, 79 Shells, buckling, 57

II. Non-linear dynamic instabilities